Outlines of Paint Technology

W. M. Morgans

Ph.D., C.Chem., F.R.S.C.

Consultant in Paint Technology

In Two Volumes

VOLUME 1: MATERIALS

18 20

Quis separabit nos

CHARLES GRIFFIN & COMPANY LTD

London and High Wycombe

CHARLES GRIFFIN & COMPANY LIMITED
Copyright © 1982

Registered Office:
Charles Griffin House, Crendon Street
High Wycombe, Bucks HP13 6LE
England

First published (in one volume) 1969
Second edition (2 volumes):
    Volume 1   1982

British Library Cataloguing in Publication Data
Morgans, W. M.
    Outlines of paint technology. – 2nd ed.
    Vol. 1: Materials
    1. Paint
    I. Title
    667'.6   TP936
    ISBN 0-85264-259-8

Filmset in Great Britain by Latimer Trend & Company Ltd, Plymouth
and printed by Henry Ling Ltd.

Preface to Volume 1

The object of this work remains essentially as in the first edition, namely, to provide an "Outline" of the subject and so provide the reader with a general framework which can be filled in, if desired, by a study of more specialized works and original papers. The book makes no claim to be a comprehensive treatise, and the number of examples quoted as illustrations in each class is, in general, but a fraction of the total.

When considering this revision, it became apparent that several advantages would accrue from a division of the work into two volumes, Volume One to cover Materials and Volume Two, Finished Products. Volume One is likely to be of interest to technologists in other industries such as printing inks and artists' colours to whom Volume Two is unlikely to make a great appeal. Other factors in favour of two volumes are ease of handling and division of expenditure.

Advances made in paint technology since the date of publication of the first edition have necessitated a number of changes to the text. In addition, the increased awareness of the health hazards associated with the use of certain paint ingredients has resulted in several Acts and Regulations to which reference is made in the text.

The manufacture of pigments is now a highly specialized industry and so the description of manufacturing operations and plant has been reduced, more attention being paid to properties and performance. Colour science has become more important and this is reflected in the text. Some pigments (and resins) which are no longer used but have some historic interest are listed in Appendix E.

Oleoresinous varnishes based on phenolic or modified phenolic resins continue to be used on a limited scale, but those based on fossil resins are virtually obsolete in this country. The treatment of these, therefore, has been curtailed and more attention paid to the wide range of alkyds and polymeric binders generally. Increased space has also been given to water-based media which are growing in importance for reasons of low toxicity, non-pollution of the atmosphere, and absence of fire risk. Oil-bound distempers, which have been replaced by latex emulsions, have been deleted. The section on water-soluble media has been extended to include media for both anodic and cathodic methods of electrodeposition.

The author extends his thanks to many friends and colleagues for helpful discussions and suggestions and to the following companies for their ready assistance: BASF (U.K.) Ltd., Coulter Electronics Ltd., Degussa, ECC

International Ltd., Pye-Unicam Ltd., Tintometer (Sales) Ltd., and Tioxide International Ltd. To those who kindly allowed reproduction of illustrations, acknowledgement is made with each caption.

A special tribute is due to my wife for her patience and understanding and to Mrs Audrey Walford who so efficiently deciphered my notes and typed the manuscript.

Finally, the author's thanks are accorded to the publishers for their patience and help at all times.

W. M. MORGANS

Walberswick, May 1981

Contents

ABBREVIATIONS OF JOURNALS

Brit. Chem. Eng.	British Chemical Engineer
Chem. & Process Eng.	Chemical and Process Engineering
Chem. Brit.	Chemistry in Britain
Chem. Rev.	Chemical Review
Ind. Eng. Chem.	Industrial and Engineering Chemistry
J. Am. Chem. Soc.	Journal of the American Chemical Society
J. Appl. Chem.	Journal of Applied Chemistry
J. Appl. Chem. Abs.	Journal of Applied Chemistry Abstracts
J. Appl. Pol. Sci.	Journal of Applied Polymer Science
J. Chem. Soc.	Journal of the Chemical Society
J. Coatings Tech.	Journal of Coatings Technology
JOCCA	Journal of the Oil and Colour Chemists' Association
J. Paint Tech.	Journal of Paint Technology (now Journal of Coatings Technology)
J. Sci. Inst.	Journal of Scientific Instruments
JSCI	Journal of the Society of Chemical Industry
Off. Dig.	Official Digest (now Journal of Coatings Technology)
Pig. & Resin Tech.	Pigment and Resin Technology
POCJ	Paint, Oil and Colour Journal (now Polymer Paint and Colour Journal)
PPCJ	Polymer Paint and Colour Journal
Trans. Far. Soc.	Transactions of the Faraday Society
Zeit. Tech. Physik.	Zeitschrift für Technische Physik

1 Introduction

The nature of paint

The major ingredients of paint are pigments (including extenders), binder (which may be organic or inorganic), and solvent or thinner. A dispersion of the pigment in the binder constitutes the paint film, the properties of which are determined to a large extent by the nature of the binder. The solvent or thinner is used to render the pigment/binder mixture sufficiently fluid for application as a thin film, after which it is lost by evaporation.

Small quantities of other materials are incorporated, depending on the type of paint and the purpose for which it is intended. Conventional air-drying paints contain "driers" which are organic salts of the metals lead, cobalt, zirconium and manganese. Generally two, or sometimes three of these are used to obtain the optimum drying and hardening of a paint film (see Chapter 10). Organic salts of calcium and zinc are used for special purposes. In addition, anti-skinning agents, anti-settling agents, thixotropic agents, or fungicides may be incorporated.

Historical

Paints have been used for decorative purposes for many centuries. The paintings produced in ancient Egypt have been shown to incorporate gum arabic, gelatin, egg white and beeswax. Gum arabic was also used by the Persians.

In classical Greece extensive use was made of paint in sculpture (for the hair, lips and eyes of statues), architecture, and in painting ships. It was also used in interior decoration and by artists.

The evolution and use of paints in Europe was mainly by artists, although a limited amount was consumed in ship painting. The Industrial Revolution, however, created a demand for paint to protect machinery and this was the start of the modern paint industry.

The paints were based mainly on drying oils, and this type remained in common use to about the end of the first quarter of the present century when oleoresinous varnishes and, later, alkyd resins gradually replaced the oils. Nevertheless, British Standard specifications for oil-type paints are still current.

Since World War II, rapid developments have taken place in the field of high polymers, leading to new types of resin suitable for use in paints. These have enabled the paint technologists to satisfy the demand for high-performance coatings with rapid drying or curing times to meet modern production schedules. Such resins include epoxies, polyurethanes, alkyd/amino, vinyls, acrylics, silicones, and chlorinated rubber.

The use of cellulose nitrate ("nitrocellulose") as a film-forming material dates from the latter part of last century, but it came into more general use for certain types of finishing only early in this century. As a non-convertible coating, more than one coat was difficult to apply by brush, and a great impetus was given to its use in industrial work by the introduction of the spray gun. The rapid drying, high gloss, and durability were found to provide the ideal finish for motor cars and this remained the main use for many years. Another application of cellulose nitrate lacquers was in the finishing of furniture.

In the car industry, cellulose nitrate has now been relegated to re-finishing, whilst for furniture it has encountered serious opposition from polyester and polyurethane finishes.

The general nature and functions of the major paint ingredients are as follows.

Pigments

These are finely divided solids, the average particle size of which can vary from 0·2 to 20 μm (micrometres). They may be inorganic or organic in constitution. The general classification of pigments can be illustrated by the following scheme —

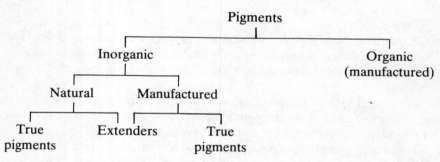

Extenders have been included with the inorganic pigments since they are all inorganic solids. They differ from "true" pigments in their behaviour when dispersed in organic media. True pigments exhibit opacity or hiding power in varying degree, whereas extenders are practically transparent. Extenders are used in certain types of paints (notably undercoats, primers and some low-gloss finishes) to modify or control physical properties. They make no contribution to colour (unless they are very impure) or to opacity.

True pigments are used to provide colour and opacity or hiding power. In finishes they contribute to durability. A pigmented film is more weather-resistant than an unpigmented film of the same binder. In primers for metals, specific pigments are used to check or inhibit corrosion of the metal.

The majority of natural pigments are oxides or hydroxides of iron but may contain appreciable quantities of clay or siliceous matter. The colours are less bright than the corresponding manufactured oxides and hydroxides.

The manufactured inorganic pigments contain the whites and a wide range of colours, including yellows, reds, oranges, greens and blues. Carbon black, consisting essentially of elementary carbon, is usually included in the inorganic pigments.

The organic pigments cover the entire spectrum range, but the brilliance and opacity vary considerably. There are no white organic pigments and organic blacks find only limited use; carbon black satisfies most paint requirements.

In general, organic pigments are brighter than the inorganic counterparts but show much greater variation in opacity and in lightfastness, the latter particularly when mixed with white. Organic pigments derived from plant and animal sources are no longer employed in paint manufacture. Although some possess bright self-colours, they lack permanence. A number which are of historic interest are listed in Appendix E.

The chemical constitution of pigments (and dyes) is set out in the Colour Index [1] and in the following pages the Colour Index description (CI —) follows the pigment name.

The testing of pigments is usually carried out by the generally accepted methods set out in British Standard 3483 [2] to which frequent reference is made in the text.

Following the introduction of the Health and Safety at Work Act, 1974, greater attention is now paid to the avoidance of any toxic or irritant effects arising from dusts created during the use of pigments and extenders (as well as solvent vapours). This matter is discussed under "Toxicity" in Chapter 2.

Binder (or film-former)

This is the continuous phase in a paint film and is largely responsible for the protective and general mechanical properties of the film. Film properties depend also on the nature of the pigment, the degree to which the pigment is dispersed in the binder, and the volume of film occupied by the pigment (the "pigment–volume concentration").

The majority of binders are organic materials—oleoresinous varnishes, resins containing fatty acids from natural oils (alkyds, epoxy esters, urethane oils), treated natural products (cellulose nitrate, chlorinated rubber), and completely synthetic polymers.

A few inorganic binders are used, notably pre-hydrolysed ethyl silicate, quaternary ammonium silicate and alkali silicates (sodium and lithium) which are pigmented with zinc dust to give primers for steelwork. The two latter silicates are also used in fire-resisting paints for stage scenery.

Organic binders can be divided into two general classes—convertible and non-convertible.

Convertible binders

These undergo a chemical reaction in the film. In the conventional oxidizing binders, oxygen is absorbed and the film slowly sets and then dries to a product which is no longer soluble in the solvents used in the liquid paint. The drying and hardening of the film is catalysed by "driers" and since oxygen absorption is involved, there is a limit to the thickness of film which will dry through.

The two-pack materials, notably epoxies and polyurethanes, cure by chemical reaction between two components in the film. Absorption of oxygen is not involved and thicker coatings can be applied. By the use of low molecular-weight resins and hardeners, "solventless" epoxy systems can be produced and very thick films can be applied and cured.

Thermosetting binders

These are also known as "thermohardening" and as "stoving" types. As in other convertible coatings, film hardening is the result of three-dimensional linking of the polymer molecules, but in this case it is brought about by the action of heat. The film is cured much more rapidly and attains the full resistance to reagents and solvents in a fraction of the time required by the air-drying types.

Stoving of the films is carried out in either hot-air convection ovens or in infra-red ovens heated by either electricity or gas. Certain types of coating, especially those which cure by a free radical mechanism, can be cured in seconds only, by either an electron beam or ultraviolet radiation. Since no heat is involved this process is suited to wood finishes and is used for polyesters and acrylics.

Non-convertible binders

These do not depend on chemical reaction of any sort for film formation, the process consisting solely of evaporation of solvents. The film remains soluble in the parent solvent blend and is therefore classed as "non-convertible". Such materials are often designated "lacquers" and are usually applied by spray (especially a second coat) since the solvents present tend to dissolve the previously applied coat. This would be disturbed or "lifted" if brushing were employed. Typical examples of non-convertible binders are chlorinated rubber and cellulose nitrate.

A very large number of polymers are now available, and by the choice of suitable combinations it is possible to produce industrial finishes capable of withstanding almost all types of conditions and service.

Solvents (or thinners)

These are usually volatile organic liquids in which the binder or film former is soluble. When the paint film is deposited on a surface the solvent should evaporate completely. It plays no part in film formation and is used solely as a means of conveying the pigment/binder mixture to the surface as a thin uniform film.

Classes of materials used as solvents include aliphatic and aromatic hydrocarbons (with the exception of benzene), esters, ketones, esters and ether–esters of ethylene glycol, and alcohols.

The fact that these materials are volatile, flammable and possess a degree of toxicity has raised a number of serious problems. Atmospheric pollution by solvents can be appreciable in areas containing a number of paint-consuming industries. The problem was particularly acute in Los Angeles and resulted in the "Rule 66" legislation [3] designed to control the emission of solvent vapours into the air.

Organic solvents can be taken into the human body by swallowing (this is uncommon), by the respiratory system, or by absorption through the skin. The effects can vary from skin irritation to damage to internal organs. A great deal of legislation, including the Health and Safety at Work Act, 1974, now exists setting out the precautions to be taken in handling solvents and paints. This matter is discussed in more detail in Chapters 2 and 9.

A second hazard associated with most organic solvents is that of fire and explosion and this also is discussed in Chapter 9.

References to Chapter 1

[1] Published by the Society of Dyers & Colourists, Bradford.
[2] BS 3483:1974. Methods for testing pigments for paints. British Standards Institution, 2 Park Street, London, W1A 2BS.
[3] LARSON, E. C. & SUPPLE, H. E., Los Angeles Rule 66 and exempt solvents. *J. Paint Tech.*, 1967, **39**, No. 508, 258.

2 Pigments: general physical properties

A pigment can be defined as a finely divided powder which can be dispersed in media of various types to produce paints or inks. It is insoluble in the medium and confers the following properties on the mixture:

(a) *Colour*, for decorative effect or aesthetic appeal.
(b) *Obliteration or hiding power*. It should obliterate the surface, whether it is metal, wood, stone, or an earlier coat of paint. As will appear later, pigments differ considerably in their hiding power and, in some instances, several coats of paint may be necessary to effect complete obliteration of the surface.
(c) *Protective properties*. These are important in paints on exposed wood and certain types of metals, particularly iron and steel. A suitably pigmented varnish is more weather-resistant and therefore more protective than the unpigmented material.
(d) *Corrosion repression*. Primers for iron and steelwork contain special pigments chosen for their corrosion-repressing properties.

A special class of pigment, usually described as "extender", generally possesses little or no colour or hiding power. Such materials are often included in paints to modify physical characteristics such as gloss and flow properties.

Classification of pigments

Pigments may be inorganic or organic in composition. The inorganic types may be processed ores or may be manufactured, but the organic pigments in present-day use are all manufactured. Many earlier organic types were obtained from vegetable or animal sources, but these have been replaced by the manufactured pigments which are more consistent and possess greater all-round stability. Since many of the natural colouring-matters are of historic interest, a selection is listed in Appendix E.

Physical properties

A number of physical properties are common to all pigments and are described in the following pages. Chemical properties are more specific and will be considered under the pigment classes or individual pigments.

6

COLOUR

Colour can be defined as a sensation resulting from activation of the retina of the eye by electromagnetic vibrations which we describe as light-waves. These light-waves differ in wavelength and each wavelength produces a specific sensation. Thus the longest waves produce the sensation which we describe as red and the shortest give the sensation of blue. The range of sensations between red and blue, which we call colours, is known as the *visible spectrum*. The fact that daylight contains the whole visible spectrum is shown by the simple experiment of passing a narrow beam of white light through a glass prism. The beam is opened out and a spectrum is produced.

The relationship between colour and wavelength is illustrated in the following diagram.

Relationship between colour and wavelength

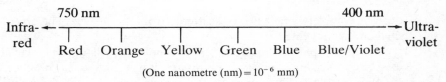

(One nanometre (nm) $= 10^{-6}$ mm)

The visible spectrum consists of only a part of the whole electromagnetic range of wavelengths or vibrations. Beyond the red are the infra-red (heat) rays and the much longer radio waves. Wavelengths shorter than the blue/violet are the invisible but intensely reactive ultraviolet, X-rays, and gamma rays.

Primary colours

By the use of mixtures of narrow wavebands of three spectrum colours—red, green and blue—it is possible to produce white light as well as a very close approximation to the range of spectrum colours. The three spectrum colours are known as the "light primaries" (or, sometimes, "additive primaries") and correspond approximately to wavelengths of 650 nm, 530 nm and 460 nm.

Additive colour mixing

The light primaries can be combined by directing the light from three lanterns, each fitted with the appropriate filter to transmit the desired wavelength, on to a white surface. The reflected colour is a mixture of the three incident beams and no appreciable energy is lost; the mixing is therefore said to be "additive". The matching or measurement of colours by mixtures of light primaries is the basis of colorimetry (page 11).

Complementary colours

When the three primary colours are mixed in pairs, red and blue give magenta, red and green give yellow, green and blue give cyan. If each of these resulting colours is mixed with the remaining primary, the result is white light, i.e. magenta + green, yellow + blue, red + cyan are pairs of colours which can produce white light on mixing. Such pairs are known as complementary colours or, sometimes, as complementaries.

Another, more practical, way of describing complementary colours is by considering a coloured material such as a pigment. When illuminated by white light, the reflected wavelengths (the observed colour) are complementary to those absorbed.

Magenta, yellow and cyan are sometimes known as "subtractive primaries".

Colour of objects

When white light falls on a solid object it can be affected in one or more of the following ways:

(i) It can be completely reflected or scattered from the surface. The object will then appear white.

(ii) It can be completely absorbed and the object will appear black. Complete reflectance or absorption are not achieved in practice, although certain surfaces such as magnesium oxide and high-quality carbon black, respectively, are close approximations.

(iii) Some of the wavelengths can be absorbed and some reflected. The object will then appear coloured, the colour being determined by the nature of the reflected wavelengths.

In practice, (iii) is often accompanied by a certain amount of (i) or (ii). If accompanied by (i), that is, by some reflectance over the whole spectrum (white), the colour will appear lighter. When accompanied by (ii) or absorption over the whole spectrum (black), the colour will appear darker or often dirtier.

Subtractive colour mixing

When a mixture of coloured pigments is exposed to white light, the colour observed depends on the absorption and reflectance characteristics of the individual pigments. Consider, for example, a mixture of yellow and blue pigments. The blue absorbs the red, orange and yellow wavelengths whilst the yellow absorbs the blue and violet. The mixture absorbs all the spectrum colours except green, which is the resulting colour. The green is produced by subtractive colour-mixing. (This is an ideal condition—in practice the absorption bands are not clearly differentiated and overlap occurs. There are, for instance, "reddish" and "greenish" blues, "reddish" and "greenish" yellows; and a range of greens will be produced from the combinations.)

Again, a mixture of red and green pigments, unlike the light primaries, will not produce yellow on account of absorption. Yellow is therefore introduced as a primary in pigment mixing, and the pigment primaries are red, yellow and blue. However, it is not possible to produce white by any combination of these, but, as a result of absorption effects, a range of greys is obtained.

Colour perception

On entering the eye, rays of light pass through the transparent lens and are focused on the interior surface at the back of the eye. This surface comprises the sensitive ends of the optic nerve and is known as the "retina". The wavelengths of the incident light produce characteristic impulses which pass to the brain by the optic nerve and produce the sensations associated with colour.

The ends of the optic nerve form a large number of *rods* and *cones* which have distinct functions. The rods do not react to colour but are sensitive to differences in intensity. They are responsible for night vision, when objects appear as different intensities of dark grey or black.

The cones are sensitive to colour but do not function efficiently in poor light. There is an old theory, put forward in the nineteenth century by Young and Helmholtz, in which cones are classified in three types, i.e. (i) those sensitive to long wavelengths which produce the sensation of red, (ii) those sensitive to medium wavelengths which produce the sensation of green, and (iii) those sensitive to short wavelengths which produce the sensation of violet. This theory has been difficult to prove but fits in with modern ideas of colour-blindness.

The sensitivity of the average human eye varies with the wavelength of the light. It is greatest in the green waveband (about 560 nm) and decreases rapidly on each side (Fig. 2.1).

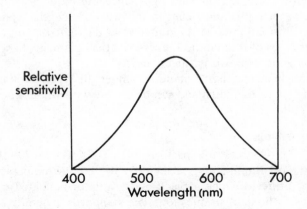

Fig. 2.1 Sensitivity of the human eye

Colour-blindness

A person with normal colour vision can match any colour of the visible spectrum by using red, green, and blue lights in the correct proportions. Such a normal person is known as a *trichromat*.

Colour-blindness in females is comparatively rare, but with men a common type involves difficulty with one of the primaries, so that the colour match produced will be different from that of a trichromat. About 1 percent of males have difficulty with red and about 4·6 percent with green. Difficulty with blue is uncommon. Persons suffering from this defect are known as *anomalous trichromats*.

Dichromats are completely blind to one of the primaries and would use only two for a colour matching. The most common forms are blindness to red or green. Blindness to blue is very rare.

Totally colour-blind people, *monochromats*, are extremely rare. They can distinguish only light and dark, and their world, therefore, is one of varying shades of grey.

It is possible to diagnose deficiencies in colour vision by the use of the Ishihara Colour Chart [1], or the Colour Matching Aptitude Test [2].

Colour matching

This involves the mixing of suitable pigments (or paints) to match a given standard and requires (i) an operator who has normal colour vision, and (ii) a standard source of illumination. The latter is important for comparison of the standard and the matching; and whenever possible, in the Northern hemisphere, a north light is used. In the absence of a north light it is possible to use a source of illumination in which the wavelengths in white light have been simulated by blending the light of fluorescent and tungsten lamps. These are often called *daylight lamps*.

The colour of an object illuminated with white light can be different when viewed in, for example, tungsten light since the latter is deficient in the blue/violet wavelengths and richer in the red wavelengths, compared with white light. Two colours can be matched in daylight but may not match under other types of illumination unless identical pigments have been used. This phenomenon is known as "metamerism".

A colour which matches a standard under all types of illumination is known as a "spectral" match and can be obtained only if identical pigments are used in both.

Illuminant standards

Standard light-sources are specified in the CIE system (see below) and called "Standard Illuminants". They refer to "colour temperature" which is the temperature to which a black-body radiator is heated to emit light which matches the light in question. The standards generally adopted are the following:

Illuminant A,   colour temperature 2850°K, corresponds to the light from a tungsten filament lamp.

Illuminant B,   colour temperature 4800°K, corresponds to noon sunlight.

Illuminant C,   colour temperature 6740°K, corresponds to north sky light or average daylight. This is the most frequently used.

Illuminant D_{65},   colour temperature 6500°K, is a more recent addition to the range. It includes more ultra-violet radiation than does Illuminant C and covers all wavelengths between 300 and 830 nm. Like Illuminant C, it is a close approximation to average daylight.

COLOUR MEASUREMENT

Instruments for the measurement of colour fall into two general classes:

(i) Colorimeters, in which the colour is measured in terms of three primaries.

(ii) Spectrophotometers, which measure the reflectance at a number of wavelengths over the entire spectrum.

Colorimeters

The basic principle of these instruments is the evaluation of a colour by matching with a mixture of three additive primaries (usually visually) or by measuring the reflectance of the three primaries from the illuminated coloured surface by photoelectric methods. There is also a colorimeter (Lovibond–Schofield Tintometer, page 12) which uses the subtractive primaries.

Additive visual colorimeter

An instrument which has been extensively used for colour measurement for a number of years is the Donaldson Colorimeter [3]. The coloured surface is illuminated with white light, and the reflected beam, taken to avoid specular reflection, is directed on to one half of a viewing screen. Beams of the three primary colours, produced by appropriate filters, are directed on to a matt white surface adjacent to the specimen and the resulting reflected mixture is passed into an "integrating sphere". The interior wall of the sphere has a matt white surface which gives effective mixing of the beams by repeated diffuse reflectance. The relative amounts of the three primaries can be varied until a visual match for the specimen is obtained. The figures so obtained are known as "tristimulus values".

Although visual colorimeters are the least expensive and continue to be used to some extent, they have been largely replaced by photoelectric instruments in which errors due to variations in operators' colour vision are eliminated.

Photo-electric colorimeters

These operate on one of two basic principles:

(a) The coloured surface is illuminated by each of the three primaries in turn, the intensity of the reflected beam being measured by photoelectric cell and galvanometer. A matt white surface is used as a reference standard.

(b) The coloured surface is illuminated by a standard source, and the reflected beam is passed through red, green and blue filters in turn. The intensities of the transmitted beams are measured as in (a) and the results recorded as tristimulus values.

Although the optical and electrical systems in (b) are less simple than in (a), a number of commercial instruments operate on this principle. Two of the most widely used are the Colormaster Differential Colorimeter [4] and the Color-Eye [5].

In the Colormaster instrument, a beam of light from a standard source is split into two, one of which illuminates the standard white surface and the other the coloured specimen. The reflected beam from the latter is passed through the red, green and blue filters in turn and the tristimulus values determined.

This colorimeter is particularly suited to measuring small colour-differences. Metamerism can also be detected and measured by changing the light source.

The Color-Eye embodies a large integrating sphere to which the coloured and white surfaces are attached and by which they are illuminated from a standard source. Light from the coloured surface is passed through the three primary filters in turn to give the tristimulus values. The Color-Eye also embodies a number of colour filters enabling it to function as an abridged spectrophotometer.

The *Fibre Optics Colorimeter*, designed by the Paint Research Association [6], enables colour measurements to be made on surfaces which cannot be placed in or brought to an instrument. It utilizes the principle that light is conducted along a glass fibre or rod. A "sensing head" is connected to a "detector head" by a flexible lead consisting of a bundle of glass fibres. The fibres are surrounded by protective material of lower refractive index to prevent loss of light energy from the sides of the fibres. Light reflected from the coloured surface is picked up by the sensing head and passes to the detector head from which it passes through the three primary filters in turn to give the tristimulus values.

Subtractive colorimeters

The best-known instrument of this type is the Lovibond–Schofield Tintometer [7]. The coloured specimen is illuminated by a standard source, and the reflected light is directed to one half of the field of view of the eye-piece. Light from a similar source is reflected from a white magnesium carbonate block placed adjacent to the specimen, and passes through a case which contains a number of sliding racks of permanent glass filters. The colours employed are the subtractive primaries, yellow, cyan and magenta. The series are inter-related in such a way that when equal values of the three primaries are superimposed and viewed through the eyepiece, a grey or neutral colour is observed as a result of absorption. The principle is illustrated in Fig. 2.2. White light is passed through the three coloured filters in turn, the filters being designed to absorb over about one-third of the spectrum and to transmit it over about two-thirds. Ideally, the result depicted in Fig. 2.2 would be obtained, but in practice filters of varying intensity are employed, so that no filter absorbs completely.

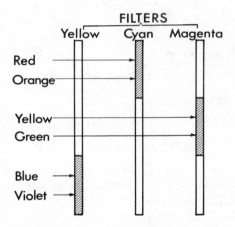

Fig. 2.2   Principle of the subtractive colorimeter

The Lovibond filter racks each contain nine filters of the Lovibond Colour Scale, and as the rack is moved along, each filter is brought in turn into the path of the white light. For each of the primaries the racks are numbered 0·1–0·9, 1·0 to 9·0 and 10 upwards, according to requirements. Thus by interposing suitable filters in the path of the white light so that the two halves of the eyepiece are matched, it is possible to specify the colour in terms of Lovibond values. These can be recorded as such, or can be converted to CIE units (described later).

An additional refinement consists of a device whereby the brightness of either half of the field of view can be adjusted.

When the tintometer is used for the measurement of the colour of liquids, the sample to be tested is placed in a rectangular cell in the path of light reflected from a second magnesium carbonate block. The colour is then matched in the manner described previously.

Reflectance spectrophotometers

If, instead of measuring reflectance values at three narrow wave-bands as in colorimetry, a large number of reflectance values can be determined covering the entire spectrum, then a more complete picture of the reflectance characteristics of a colour can be obtained. If these reflectance values are plotted against wavelength a "spectral reflectance curve" (Fig. 2.3) is obtained which is a permanent record of the reflectance characteristics of a coloured surface or pigment.

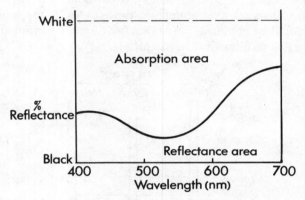

Fig. 2.3   Spectral reflectance curve

The general principle of a spectrophotometer is illustrated in Fig. 2.4, but commercial instruments can vary appreciably in design [8]. A narrow beam of white light is split into a spectrum by a prism or diffraction grating and the spectrum thrown on to a screen. The screen embodies a movable slit by means of which a very narrow band of wavelengths is allowed to illuminate the standard white surface and the colour to be measured. The intensity of the reflected beams is measured by photocell and a special type galvanometer. The figure obtained is the reflectance of the particular waveband from the coloured surface in relation to the reflectance from the white surface. Any

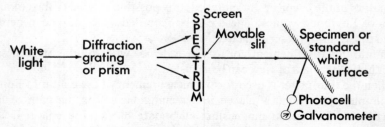

Fig. 2.4   Principle of the spectrophotometer

number of measurements over the spectrum can be made in this way, but the process of measurement and recording can be time-consuming and tedious.

To overcome this drawback, automatic recording spectrophotometers have been designed in which a spectral reflectance curve is produced on graph paper. A well-known instrument is the Hardy recording spectrophotometer [9] which produces a spectral reflectance curve very rapidly. An attachment can also give the CIE chromaticity values. Among other recording instruments which are widely used are the Pye-Unicam SP8–100 [10] which also can be fitted with an addition to give tristimulus values and CIE chromaticity co-ordinates, the Beckman Model 26 [11] and the Hunterlab D54P–5 [12]. The latter can also be adapted for production of tristimulus values and chromaticity co-ordinates.

Abridged spectrophotometers are simpler and less sophisticated instruments and produce a spectral reflectance curve for a limited number of measurements. Although less accurate than the full spectrophotometer, they are often adequate for a number of control and routine measurements. They incorporate nine to sixteen narrow-band filters mounted round a wheel through which the sample and the standard magnesium carbonate can be illuminated. The filters are used in turn to illuminate the magnesium carbonate and then the specimen. The reflected beams are directed to a photoelectric cell and galvanometer. The figures obtained give the percentage of the particular wavelength reflected from the specimen relative to that reflected from the white surface.

Spectral reflectance curves

These reveal more about a coloured surface than do colorimetric measurements. The state of a curve can give an indication of the type of pigment present and so provide guidance in establishing a match. Spectral matches will give identical spectral reflectance curves, but if metamerism is present the curves usually cross each other two or three times. Another important use of these curves is in calculation of absorption and scatter coefficients used in colour-mixture formulation and control.

Use of spectral reflectance curves in colour-matching

Kubelka and Munk [13], as a result of their studies on the optical properties of dispersed powders, derived a relationship between the reflectance of a pigmented film and the absorption and scatter coefficients of the pigments. This is now referred to as the Kubelka–Munk (or KB) equation of which the following is a simplified form:

$$\frac{K}{S} = \frac{(1-R)^2}{2R}$$

where K = coefficient of absorption,
$\quad\quad S$ = coefficient of scatter,
$\quad\quad R$ = reflectance of film at maximum opacity, i.e. when the opacity no
$\quad\quad\quad\quad$ longer increases with the thickness.

For opaque pigmented films the expression is valid at any wavelength in the visible spectrum.

Values of R can be taken out at intervals from a spectral reflectance curve and from these the corresponding K/S values can be calculated. With mixtures of pigments the K/S values are additive, and for pale shades, where absorption is small in relation to scatter, the values can be programmed for an analogue computer. By this means it is possible to determine the quantity of individual pigments required to match a given pale colour or to correct a shade.

The calculations are more involved for deep or saturated colours and the original Kubelka–Munk equations may be required with a digital computer.

COLOUR SYSTEMS AND ATLASES

The CIE System [14]

This is an internationally agreed system of colour specification which was agreed in 1931 by the Commission Internationale de l'Eclairage, a body consisting of a number of international organizations. The essence of the system is the definition of colour in terms which are independent of a particular individual. It assigns numerical values to the sensations experienced by an average individual in looking at colour.

According to the CIE system, colour possesses three subjective attributes — hue, saturation, and lightness or luminosity. These correspond to the objective variables dominant wavelength, purity, and luminance. The latter represents the intensity of light reflected from a coloured surface relative to that reflected from a white surface.

In observing a coloured object, three factors are involved. These are (a) the nature of the light-source or illuminant, (b) the visual response of the observer, and (c) the reflectance properties of the object. These are specified as follows:

(a) *The light source or illuminant.* Three sources were specified originally and called "Standard Illuminants". A further source (D_{65}) was introduced more recently. These have been described on pages 10–11.
(b) *The visual response of the observer.* The system defines a "standard observer" as one whose response to the three primaries is the average of a large number of observers examined.
(c) *Reflectance properties of the object.*

In the practical matching of certain spectral colours using the three additive primaries, difficulties have arisen due to the fact that mixtures of the primaries are less saturated than the spectral colour to be matched. The latter requires "desaturating" and this introduces difficulties. To overcome this problem the CIE system has postulated three "unreal" primaries designated X, Y and Z which can be calculated from the tristimulus values determined colorimetrically. They are also known as CIE tristimulus values and are indicated directly on some modern colorimeters.

In order to represent a colour on a two-dimensional chart, the CIE tristimulus values are converted to chromaticity co-ordinates in the following way:

$$x = \frac{X}{X+Y+Z}, \qquad y = \frac{Y}{X+Y+Z}, \qquad z = \frac{Z}{X+Y+Z}$$

Since $x+y+z=1$, only two of these are required to locate a colour on a two-dimensional graph and are designated x and y. A third co-ordinate is the "luminance factor", designated Y, the value of which is related to the spectral response of the human eye when observing the colour.

The chromaticity diagram

Guild and Wright determined the tristimulus values at a large number of wavelengths over the entire spectrum and converted the figures to CIE chromaticity co-ordinates. When these were plotted the curve illustrated in Fig. 2.5 was obtained. The curved line is known as the "spectrum locus" and extends from blue (400 nm) to red (700 nm). The straight line joining the red and violet represents the non-spectral purples. All real colours are contained within the closed area.

Saturation or purity of a colour Point C indicates the position of standard illuminant C (or white); suppose D is the position of a colour as given by the chromaticity co-ordinates. If the line CD is extended to meet the spectrum locus at E, then this will be the dominant wavelength (or hue) of the colour. The saturation or purity of the colour is given by CD divided by CE. The complete CIE specification for a colour is then given by the chromaticity co-ordinates, the luminance factor Y (which is not shown in the diagram but which would be perpendicular to the surface), and the nature of the illuminant (A, B, C or D_{65}) — C and D_{65} being most commonly used.

Applications of the CIE system

A CIE specification gives a permanent record of a colour in numerical values which relate to the actual sensations experienced by an average observer in viewing the colour. It is being used for the international standardization of colours for specific purposes, e.g. transport signs.

It is used to record colour changes such as fading or darkening and, in production, as a guide to colour correction.

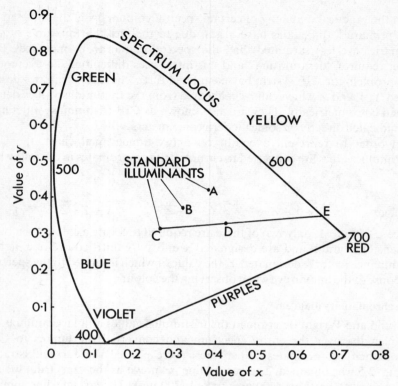

Fig. 2.5   Chromaticity diagram

The Munsell system

This comprises a colour atlas which was devised by Munsell, who assigned three subjective variables to each colour, namely *hue* (basic colour corresponding to dominant wavelength), *value* (lightness), and *chroma* (purity or saturation). He envisaged (see Fig. 2.6) all colours as contained in a solid having a vertical axis with white at the top and black at the bottom. This axis or trunk is divided into nine blocks which represent values (grey content).

On a horizontal circumference surrounding the axis are placed ten principal colours or hues which are designated by their initial letters. These hues are Red, Red/Purple, Purple, Purple/Blue, Blue, Blue/Green, Green, Green/Yellow, Yellow, and Yellow/Red. Each hue is divided into ten, with the pure colour at 5 and the positions of the hues as illustrated in Fig. 2.7. Points on lines connecting the hues and the axis will represent purity or chroma. These are shown as blocks and are numbered in six steps of 2 (2 to 12) from the axis.

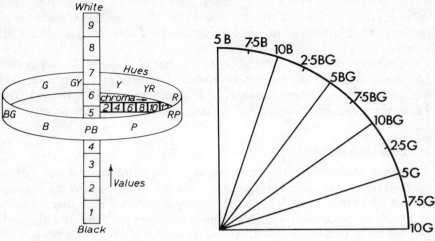

Fig. 2.6   The Munsell "Tree"   Fig. 2.7

If the solid is cut vertically to the axis in very thin sections, each section will represent all possible mixtures of a particular hue with black and white. These sections comprise the pages in the Munsell book, a page being represented by Fig. 2.8.

The complete Munsell designation of a colour is given by the sequence hue, value and chroma. For example, a colour designated 7·5 G 4/6 would be a yellowish green (hue 7·5 G) with a lightness value of 4 and a chroma of 6.

It is essential in using the Munsell colour atlas that all colour comparisons be made under standard illumination.

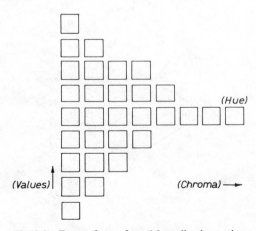

Fig. 2.8   Form of page from Munsell colour atlas

The Munsell description of a colour corresponds to the CIE attributes of dominant wavelength (hue), luminance (lightness), and saturation (purity). All Munsell colours have been assigned the corresponding CIE values.

This system possesses an advantage over that of Ostwald (*below*) in that additional hues and chroma can be added when necessary. It is now used internationally and forms the basis of British Standard 4800 (*below*). The American Standards Association also uses the system for colour specification.

The Ostwald system [3] [14]

Ostwald was particularly interested in colour harmony and this largely inspired his colour system. He arranged his *full* colours round the circumference of a circle in such a way that *complementaries* were diametrically opposite. To produce other colours his *full* colours were mixed with black and white. In making the mixtures the quantities of black and white were increased in geometric progression to give, in his view, *subjectively equal* steps. His mixings were done with sectors of colour, black, and white on a rotating disc, and as the proportions then added up to the same total, he evolved the relationship:

$$\text{Full colour} + \text{Black} + \text{White} = 1.$$

By taking a full colour and adding black and white he was able to set out the results in the form of an equilateral triangle (Fig. 2.9) with the full colour at one corner, black at the second and white at the third. This actually represents a section through a colour solid since, if the black–white axis is

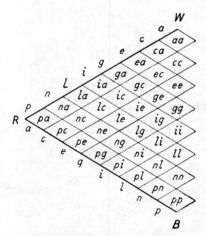

Fig. 2.9   The Ostwald triangle

arranged vertically, the series of full colours will form a circumference round the axis and the whole will be a symmetrical double cone, base to base.

The vertical axis White–Black contains the greys in blocks of definite composition. Lines drawn parallel to the White–Colour side give mixtures with equal percentages of black. Those parallel to the Black–Colour side have equal percentages of white.

In the Ostwald colour designation, a number is assigned to each full colour or hue. The white content is indicated by the letters a, c, e, g, i, l, n, and p, decreasing in this order. Black content is indicated by the same letters in the reverse order. The Ostwald designation for a colour can then be, for example, 8ge, indicating hue No. 8, with a white content g, and black content e.

Possibly as a result of its emphasis on colour harmony, the Ostwald system is popular with architects and in schools of art.

Colour cards and patterns

The need for a system of colour patterns based on scientific principles has been recognized for many years, and it is now usual for paint manufacturers in this country to use the patterns set out in BS 4800 (Paint Colours for Building Purposes) [15]. The British Standards Institution also publishes BS 381C (Colours for Special Purposes) as well as colour cards with more restricted uses, such as pipeline identification.

The British Colour Council [16] published a *Dictionary of Colour Standards* in which the colours were reproduced on several different types of material. The *Colour Index*, published by the Society of Dyers and Colourists [17] is a comprehensive work setting out the generic name, chemical constitution and properties of pigments and dyestuffs. It is used internationally.

BS 4800 contains 86 colours together with black and white. The colours are based on 12 hue positions in the Munsell scale covering the important regions and chosen so that the identity of each hue appears constant at all levels of lightness and saturation. All colours are given a Munsell reference and are described in terms of hue, greyness, and weight which correspond to hue, value, and chroma of the Munsell system.

On the card the colour patterns are arranged in blocks according to hue and are given the following numbering system: an even number indicating hue, a single letter indicating greyness, and an odd number indicating weight or intensity of colour. Examples are 08 B 15 (Magenta) and 20 D 45 (Royal Blue).

Colour comparisons of pigments

A comparison of pigment colours in the dry state is of little value as appearances often change when they are dispersed in paint media. For comparison purposes pigments are usually dispersed in acid-refined linseed oil, but other media may be used. The following is a brief summary of the method described in BS 3483, Part A1 (Comparison of colour) [18] in which either a palette knife or an automatic muller (*below*) can be used to effect the dispersion.

The dry pigment is placed on a glass palette and the oil added dropwise from a burette. The pigment is dispersed well by thorough rubbing with a stiff palette knife and the addition of oil continued until a thin paste is obtained. The comparison standard is treated in the same way. The two pastes are then applied in parallel and touching strips on a glass slide and examined under north daylight or a daylight lamp.

The colour of the undiluted pigment is often referred to as the *self colour* or *mass tone*.

The Automatic Muller

In comparing self colours and tinting strengths it is essential that the sample and standard receive exactly the same treatment so that the degree of dispersion is the same in each case. This is not easy with a palette knife but is possible by the use of the A & W Automatic Muller. In this machine the pigment is dispersed between two ground-glass plates, one of which rotates, the other being fixed. The top (fixed) plate can be loaded so that various pressures can be applied. A mechanism is provided for giving the material a prescribed number of rotations or continuous running, and an automatic cut-out is incorporated. By this means it is possible to give the same treatment to a number of samples and to obtain reproducible results.

Tinting strength (or staining power)

Coloured pigments are frequently added to whites to produce pastel shades or to other pigments to modify the colour. The weight of coloured pigment required by a given weight of white to produce a given depth of tint is an indication of the tinting strength. Tinting strengths differ widely from one pigment to another and generally increase with decreasing particle size. They are measured against a standard pigment of the same colour and composition so that tinting strengths are relative.

There is no general relationship between tinting strength and opacity. For example, carbon black has high opacity and high tinting strength, but Prussian blue, which has high tinting strength, has very poor opacity.

The tinting strength of a pigment is determined by the method described in BS 3483, Part A4 (Comparison of relative tinting strength (or equivalent colouring value) and colour on reduction in linseed stand oil using the automatic muller) [18]. The following is a brief summary of the method.

Equal weights of the standard pigment and of the sample under test are dispersed at the same concentration in the stand oil using the automatic muller. A dispersion of titanium dioxide, together with a proportion of calcium stearate and synthetic silica in a urethane-modified linseed oil, is prepared on a triple roll mill.

White paste and colour dispersions in the desired ratios (1/10, 1/20, etc.) are mixed thoroughly on the automatic muller and the colours compared on glass slides. Differences in colour intensity or hue are noted. A quantitative measure of difference in tinting strength is obtained by adjusting the ratio of white paste to colour dispersion until the same intensity is reached. Suppose x parts of test sample produce the same intensity as y parts standard pigment; then relative tinting strength $= y/x \times 100$.

The term "tinting strength" is not generally applied to white pigments which are seldom used as stainers. In this case the terms "tint resistance", "stain resistance", or "lightening power" are used. Measurement is carried out by a method similar to that for tinting strength but using a dispersed coloured pigment (generally ultramarine blue). The method is described in detail in BS 3483, Part A5 (Comparison of lightening power of white pigments) [18].

OPACITY

This is also known as "hiding" or "obliterating power" and can be defined as the ability of a pigment to hide or obliterate a surface when dispersed in a medium and spread in a uniform film over the surface.

Pigments and extenders are opaque as dry powders but show differences in opacity when dispersed in media. Pigments are opaque in both organic and aqueous media, although the degree of opacity, particularly of organic pigments, can vary considerably; extenders are practically transparent in organic media but opaque in aqueous systems.

The opacity of a pigment or extender is a function of the degree to which the incident light is scattered from the surface, and this depends on the difference in refractive index between pigment and dispersion medium. The refractive index of air is taken as 1·00. The figures for drying oils average 1·48 and for resins 1·55. The following table shows the refractive indices of some of the common white pigments and extenders.

White pigment	Refractive index
Titanium dioxide (rutile)	2·71
,, ,, (anatase)	2·55
Zinc sulphide	2·37
Zinc oxide	2·08
White lead	2·09
Extenders	
Whiting	1·58
Silica	1·55
Talc	1·49
Magnesium carbonate	1·57

The refractive indices of the white pigments all differ appreciably from those of water (1·33) and organic media and they are therefore opaque in all media. The figures for extenders, however, are extremely close to those of oils and resins and it is well known that they are virtually transparent in organic systems. They are, however, sufficiently far removed from that of water to account for their opacity in aqueous media. White pigments are generally highly reflective and scatter the incident light, but with black and coloured pigments absorption can play an important part.

The relationship between the coefficients of absorption (K) and of scatter (S) is given by the Kubelka–Munk equation (page 15). The value of S is dependent on the wavelength of the incident light, the ratio of refractive indices of pigment and medium, the particle size and particle size distribution, the degree of dispersion, and the pigment–volume concentration. The value increases with decreasing particle size, but when the latter approaches 750 nm the value of S decreases. An example of this is afforded by red iron oxide, a pigment with good opacity at the particle size normally used in paint. It can, however, be ground sufficiently fine to give a transparent film when dispersed in paint media.

Measurement of opacity

The opacity of a pigment assumes a practical significance only when the pigment is dispersed in a paint medium, and comparisons of opacities of two pigments must be done in the same medium and at the same concentration. The concentration can be expressed in terms of weight or volume; the latter is preferred if, for example, comparisons are being made between different pigments or mixtures of pigments. Although these comparisons are usually made on dry films at the same film thickness, there are occasions when comparisons of the liquid dispersions are required. Results of these comparisons may differ from those obtained on the dry films as a consequence of pigment flocculation, which sometimes takes place during drying.

Opacity of liquid paints or dispersions

This is generally determined by the Pfund cryptometer (Fig. 2.10), although rapid visual comparisons can be made using Morest charts (*below*). The base A consists of blocks of white and black glass fused together and has a linear scale engraved on the surface. The liquid paint is placed in the space between the movable plate C, of clear glass, and the metal distance-piece B. The paint is viewed from above and the wedge of paint is moved until

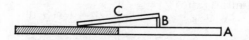

Fig. 2.10   The Pfund cryptometer (diagrammatic section)

the join in the base is just obliterated. A check reading is obtained by reversing the wedge and repeating the procedure. The thickness of paint over the line is known as the "critical thickness" and can be calculated from the scale reading and a knowledge of the wedge angle.

Opacity of dry paint films

(a) *Use of Morest hiding-power charts*

These are sheets of stiff glazed card, one square foot in area, printed in contrasting patterns of black and white, or grey and white. The following are two methods by which the charts can be used for opacity assessment.

(i) Equal weights (or volumes) of the two dispersions under test are applied uniformly to the charts and a visual comparison made when the films are dry.

(ii) The pigment dispersion is applied uniformly by brush until the contrast pattern is just obliterated. The amount of dispersion used is determined by weighing, and the opacity or hiding power of the pigment can then be expressed as sq. feet per lb, or as sq. metres per kg (multiplying by the factor 10·76).

(b) *Contrast ratio method*

In this method the pigment dispersion is applied to transparent cellulose accetate sheet in a film of known thickness. This is allowed to dry and the reflectance is measured photoelectrically over white and over black substrates. The difference is the amount of light which has penetrated the film and been absorbed by the black substrate.

The method is described in detail in BS 3900, Part D4 [19], of which the following is a brief summary.

The film on cellulose acetate sheet is produced by using steel shims 0·05 mm (0·002 in.) thick and a doctor blade. When dry the film is cut to the required size for the substrate block which comprises two blocks of glass, one black and one white, fused together. The reflectance characteristics of the blocks are standardized by comparison with magnesium carbonate.

The film is first placed on the white substrate, the opacity head placed on it, and the galvanometer set to 100. The opacity head contains a standard lamp for illumination and a photoelectric cell. The film and opacity head are then placed on the black substrate and the galvanometer reading taken. After correcting for the reflectance characteristics of the substrates, the contrast ratio is given by the equation

$$\text{Contrast ratio} (\%) = \frac{\text{Reflectance over black}}{\text{Reflectance over white}} \times 100$$

CHARACTERISTICS OF PIGMENT PARTICLES

Particle size

The particle size of a pigment exerts an influence on one or more of a number of paint properties such as gloss, opacity, freedom from settlement, and consistency.

The size, or average diameter of pigment particles, is usually expressed in micrometres (symbol μm, where μm = 0·001 mm). Pigments are never uniform in particle size but this varies over a range, with the majority of the particles approximating to one particular value and the numbers falling off rapidly on each side. Modern pigments are generally free from both coarse particles and ultra-fines. The former entail extra processing for their separation whilst the latter, if their average diameter approaches half the wavelength of light, have little or no opacity. The ultra-fines are also thought to be the cause of the *bronzing* of certain pigments, e.g. Prussian blue.

In many natural pigments and extenders the particle sizes are greater than those of prepared pigments. They are reduced by grinding and separation in mills such as the Raymond mill, or in the Micronizer. These mills are described in Chapter 3.

Determination of particle size: sieving method

Sieves are used only to separate or classify relatively coarse particles. The limits of these methods are evident from the following figures which give the aperture sizes for some British Standard sieves:

100 mesh BS sieve	Aperture	= 152 μm
240 ,, ,, ,,	,,	= 64 μm
350 ,, ,, ,,	,,	= 44 μm

Sieving methods are little used for pigments at the present time as these are supplied with a particle size range of less than 10 μm. They therefore fall into the sub-sieve range.

Particle size distribution

In view of the fact that a pigment contains a range of particle sizes it is usual to express this as a particle size distribution, that is, the percentage (by number or weight) occurring between certain limits, e.g. below 0·1 μm, 0·1 to 0·2 μm, 0·2 to 0·3 μm, and so on. These results are expressed graphically in the form of histograms which take the general form illustrated in Fig. 2.11.

A very large number of methods have been suggested for the determination of particle size distribution, but comparatively few have been adopted. Those adopted can be classified as follows.

(a) Sedimentation methods (based on the classical Stokes equation— *below*) in which the pigment is separated into a number of fractions under gravity or, more rapidly, by the use of a centrifuge. Results are usually expressed as a weight distribution.

(b) Air elutriation methods which also give a weight distribution.

(c) Counting methods using the optical or electron microscope or the Coulter counter. In these methods results are expressed as a number distribution.

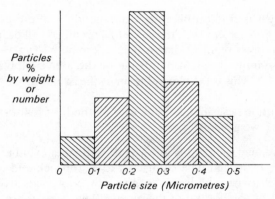

Fig. 2.11 General form of histogram

(a) *Sedimentation methods*

These are based on the principles first explained by Stokes who derived an expression relating the velocity at which a particle falls through a liquid to the size of the particle, the viscosity of the liquid, and the difference in densities of particle and liquid. The best-known form of the Stokes equation is the following:

$$v = \frac{2r^2(\rho_1 - \rho_2)g}{9\eta},$$

where v = velocity of fall of the particle
r = radius of the particle
ρ_1 = density of the particle
ρ_2 = density of the liquid
η = coefficient of viscosity of the liquid
g = gravitational constant.

It is usual to refer to pigment diameter rather than radius, so that r in the equation can be written as $d/2$. Also if v is represented as distance fallen (h) in time t, the equation becomes:

$$\frac{h}{t}=\frac{2\left(\frac{d}{2}\right)^2(\rho_1-\rho_2)g}{9\eta}$$

and this can be written in the form:

$$d=\left(\frac{18\eta h}{(\rho_1-\rho_2)gt}\right)^{\frac{1}{2}}$$

Now if h is measured in centimetres, t in seconds, η in poises, ρ_1 and ρ_2 in grams per cm^3 and g in cm/sec^2, the value of d obtained will be in centimetres. It is usual to express pigment diameter in micrometres ($\mu m = 10^{-4}$ cm), and to obtain these units the right-hand side of the above expression must be multiplied by 10^4. This is the general form of the equation set out in BS 3406 [20].

The specification describes in detail four methods of sedimentation analysis. These are of two general types:

(i) The suspended pigment is allowed to settle in a cylinder into which is fitted a pipette of special type connected to a stopcock and tap. The latter allows samples to be withdrawn at given time-intervals from a constant level indicated by the bottom of the pipette. The samples are evaporated to dryness and the weights determined. From the weights determined over a range of time-intervals the rates of sedimentation and particle sizes can be calculated, using the Stokes equation above. The well-known Andreason pipette [21] is of this type.

(ii) The suspended pigment is allowed to settle in a cylinder, as before, but the weight of settlement is determined at time-intervals, either by drawing off from the bottom, drying and weighing, or by using a sedimentation balance [22]. In the latter instrument a horizontal plate, which forms one pan of a torsion balance, is placed very close to the bottom in the settlement cylinder. The weight of sediment can then be determined at suitable time-intervals. These methods do not give the particle size distribution curve directly but involve a somewhat complex calculation.

All sedimentation methods are slow and also have a lower limit of about 0·2 μm, below which settlement is too slow. The time required for separation can be reduced very considerably by the use of a centrifuge.

The ICI–Joyce Loebl Disc centrifuge is now widely used. A suitable amount of the pigment dispersion is placed at the centre of the rotating disc and samples are removed from the periphery at known time-intervals and weighed. The method was used by Beresford [23] to determine the particle size distribution of organic pigments which, on account of their low specific gravities, settle very slowly under gravity. He found the method satisfactory for particles in the range 0·04 to 1·0 μm.

(b) *Air elutriation methods*

These methods are described in BS 3406, Part 3 [20] and the specification should be consulted for details. The general principle underlying the methods consists in the determination of the proportion of a pigment which is removed by an upward flow of a gas at a given velocity in a vertical column. The Stokes equation is used to calculate the largest size of particle removed by a gas at the given velocity.

In the determination, the gas velocity is first adjusted so that the smallest particles are carried off and collected. The velocity is then increased to remove the next fraction and the process repeated until the entire range has been separated.

(c) *Counting methods*

The optical microscope method [20]

This method consists of direct observation of a representative sample of the pigment, using a microscope fitted with a graticule. The graticule is inscribed with circles of known sizes as standards of comparison for the magnified images of the pigment particles. The number of particles in each size-class is counted and gives a size distribution by number. If the particles are of the same shape, the relative volume distribution can be calculated, and this will give a size distribution by weight if it is a simple pigment of known specific gravity.

The optical microscope has a limit of resolution of about 0·25 μm, but accurate measurement at this size is difficult. Below about 0·4 μm it is necessary to use the electron microscope.

Electron microscope method

The construction of the electron microscope and its application to the study of powders have been amply described in the literature [24, 25], and the following is a brief summary of the salient features of its application to the examination of pigments and paint films.

The great advantage of the electron microscope over the optical instrument lies in the greater resolution obtainable. The use of an electron beam combined with electromagnetic lenses having objectives of small aperture gives a great depth of focus.

Very small samples are used for examination, and the mounting of these on suitable substrates requires very special techniques, which are summarized by Herdan [26].

In the examination of pigments, information on the *surface relief* is usually required and this can be obtained either by *shadowing* the particles with a heavy metal such as gold, or, if the particles are embedded in a matrix such as a paint film, surface replicas can be made.

Shadowing

The heavy metal is deposited under high vacuum and is directed at the specimen from a low angle. Each relief feature facing the source is coated with metal but casts a shadow on the supporting substrate. No metal is deposited in these shadows. When examined in the electron beam the deposited metal appears dark as a result of electron scattering, while the shadows appear light. The photomicrographs are often printed as negatives instead of positives to obtain a realistic picture. From the length of the shadows and the angle of shadowing, the heights of the particles can be calculated.

Surface replicas

Surface replicas are prepared when the surface to be examined is in the form of a thick film or is part of a solid body.

The specimen is coated with a dilute solution of a suitable polymer and this is stripped off after drying, a negative replica of the surface being obtained. If necessary, a positive replica can then be cast from the negative and this can be shadowed and the contours examined.

Some electron photomicrographs of a few common pigments are illustrated in Fig. 4.1, 4.2, 5.1, 5.2, 7.1, 8.1.

The Coulter counter [27]

This instrument embodies an ingenious method in which size and numbers of particles are determined automatically; it is applicable to powders with a mean particle diameter of 0·6 μm to 800 μm. A very dilute dispersion of the pigment in an electrolyte is drawn through a fine orifice across which an electric current path has been established by means of two electrodes. As an individual pigment particle passes through, it displaces electrolyte in the aperture, producing a pulse proportional to its displaced volume. The pulses are amplified and appear as vertical spikes on an oscilloscope trace. They are also recorded on a digital register.

The method can be used with both aqueous and non-aqueous electrolytes, is very rapid, and can be applied to on-line quality control as well as laboratory measurements.

Surface area of pigment particles

Many paint properties, including viscosity, flow characteristics and dispersion stability, can be influenced by interaction between pigment and medium. Adsorption of certain polymer fractions and/or driers on to the pigment surface takes place frequently, and the total amount of adsorption will depend on the nature and surface area of the pigment particles. A knowledge of the latter is, therefore, of importance and the figure is commonly expressed as square metres per gram. The figure is related to the average particle size of the powder, and a rough figure for surface area can be obtained using the expression

$$\text{Surface area (m}^2\text{/gram)} = \frac{6}{D\rho},$$

where D = volume/surface mean diameter, i.e. the diameter of a particle possessing the same volume-to-surface ratio as the whole sample,

ρ = specific gravity of the pigment,

6 = factor applicable only to spherical particles; with particles of irregular shape, such as pigments, the expression can give only approximate figures.

Surface area and oil absorption

The oil absorption value of a pigment (page 33) (which represents the amount of oil required to coat the surface and to fill the voids) increases with decreasing particle size. The smaller voids resulting from closer packing of smaller particles are more than offset by the increased surface area.

Pigment particles may be smooth or irregular individuals or they may be aggregates of various sizes. These will all behave as individual particles towards oil and other media and, in the case of aggregates, the "practical" surface area will be appreciably less than the "total" surface area of the component particles. The difference depends on the type of pigment and the degree of grinding. For many manufacturing operations a knowledge of the "practical" surface area is sufficient, but in any study involving pigment surface a knowledge of "total" surface area is required.

Determination of surface area

There are two general methods in use known as the "gas adsorption" and "gas permeability" methods. A number of variations of each of these methods have been described in the literature [26].

Gas adsorption methods

The best known is the BET method [28] which consists in determining the quantity of nitrogen adsorbed by unit weight of pigment to coat the entire available surface with a layer one molecule thick. From the weight (and therefore the number of molecules) of nitrogen adsorbed and a knowledge of the cross-sectional area of the nitrogen molecule, the surface area can be calculated.

In the determination a known weight of the dried pigment is placed in an adsorption bulb, degassed under high vacuum, and the bulb and contents cooled in liquid nitrogen ($-195°C$). A known volume of nitrogen gas is admitted and the pressure noted when equilibrium is reached. The procedure is repeated with further known volumes of nitrogen. For the calculation of results, use is made of the equation of Brunauer, Emmett and Teller [28] —

$$\frac{P}{V(P_0 - P)} = \frac{1}{V_m C} + \frac{(C-1)P}{V_m C P_0}$$

where V = volume of gas adsorbed (in cm³ per gram pigment) at equilibrium
    pressure P,
P_0 = saturation vapour pressure of the adsorbate,
V_m = volume of gas (per gram pigment) when surface is coated with a
    monolayer,
C = a constant.

If $P/\{V(P_0 - P)\}$ is plotted against P/P_0 a straight line is obtained. The
values of C and V_m can be determined from the slope and intercept. From
the value of V_m and the known cross-sectional area of the nitrogen molecule
the total surface area of the pigment can be calculated.

A modification of the nitrogen adsorption method, using a Perkin–Elmer–
Shell Sorptiometer, has been described by Beresford, Carr and Lombard
[29].

Gas permeability method

The basis of this method is the establishment of a steady flow of gas
through a compacted bed of pigment. The pressure drop through the bed is
measured for several rates of flow.

The calculations are involved, and the reader who wishes to pursue the
subject further is recommended to the paper of Carmen and Malherbe [30]
or to the survey by Herdan [26].

The shapes of pigment particles

It is well known that solid materials are either amorphous (possessing no
definite or characteristic structure) or crystalline. Crystalline materials
possess a characteristic structure and shape which is retained if the particles
are reduced in size by crushing. The shapes of the crystal faces and the angles
between them can sometimes be used as a means of identification. If, however, the material has been reduced in size by the Micronizer, identification
in this way is difficult since this machine tends to remove corners and sharp
edges. Fig. 2.12 illustrates four of the basic types of crystal form.

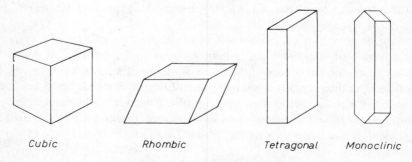

    Cubic   *Rhombic*   *Tetragonal*   *Monoclinic*

Fig. 2.12   Basic types of crystal form

A considerable number of pigments are crystalline, but in many cases the crystalline form is apparent only under high magnification as is obtained with the ultra or electron microscope. X-ray diffraction is used for fine powders. The lead chromes afford some of the best examples of crystal structure and these can usually be observed under an ordinary microscope. Thus, Primrose Chrome is rhombic, Lemon and Mid-Chromes are mono-clinic, whilst Scarlet Chrome is tetragonal. Other examples could be quoted, but these are mentioned under the individual pigments.

The shape of pigment particles influences many paint properties, including flow characteristics, tendency of the pigment to settle, the nature of settle-ment, and the durability of the paint film.

Specific gravity of pigments

According to Stokes' Law, the rate at which a solid particle settles in a liquid is proportional to the difference between the specific gravities of the solid and the liquid. This is borne out by pigments in paints, where those with high specific gravity such as red lead (sp. gr. 9·0) settle rapidly, whereas carbon black (sp. gr. 1·8) shows little or no tendency to do so. (In the case of carbon black other factors such as particle shape and nature of the surface also play an appreciable part.) A degree of thixotropy or structure is generally built into paints containing heavy pigments in order to reduce settlement.

Specific gravities of pigments are best determined by the displacement method as set out in BS 3483 [18], using either vacuum (Part B8) or centrifuge (Part B9) to remove entrained air.

The figures are used in paint formulation for calculating volume relation-ships. Pigment volume concentration (PVC), for example, controls a number of film properties, including gloss and permeability. It is calculated from the expression

$$PVC = \frac{\text{volume of pigments} \times 100}{\text{volume of pigment} + \text{volume of non-volatile binder}}$$

Many paint specifications state a minimum figure for the total volume of solid matter, a figure calculated from specific gravities. This is important in controlling film thickness or build and is used in calculating dry film thick-ness from measured figures on the freshly applied wet film.

Oil absorption of pigments

Pigments differ widely in respect of the amount of oil (or other medium) required to convert them into a workable paste. The "oil absorption" of a pigment is defined as the weight of refined linseed oil required to convert 100 parts by weight of the pigment into a smooth paste. It is determined by the method described in BS 3483, Part B7 [18] of which the following is a summary.

A known weight of the pigment is placed on a palette and acid-refined linseed oil is added dropwise from a 25 cm^3 burette. After each addition of oil the pigment and oil are well mixed with a stiff palette knife. The additions of oil and the rubbing are continued until a stiff paste of smooth consistency is obtained. The volume of oil used is converted to weight and the oil absorption calculated on 100 parts by weight of the pigment.

The oil absorption figure represents the minimum weight of oil required to coat each pigment particle and to fill the voids between them; it depends on the surface area of the pigment (which is related to particle size) and also on the nature of the pigment surface which influences the wettability.

Ease of wetting of pigments

This is of importance in the dispersion of pigments and depends on both the nature of the pigment surface and the nature of the wetting liquid.

When pigment powders are exposed to the air the particles adsorb gas and/or moisture on the surface. The extent to which this takes place and the type of material adsorbed varies from one pigment to another, and pigments can be classified as hydrophobic (water repelling) or hydrophilic (water attracting). Before a pigment can be wetted properly by an organic liquid the adsorbed gas and/or moisture must be displaced, otherwise an unstable dispersion results.

A method for comparison of the ease of dispersion of pigments is contained in BS 3483, Part B5 [18].

Surface-treated pigments

As a result of the demand for increased rates of paint production, pigment manufacturers have given considerable attention to modifying the pigment surface in order to improve wettability. This applies particularly to hydrophilic pigments which are to be dispersed in organic media. The method used consists in coating the particles with an organic compound — generally described as a surface-active agent — so that the particles assume a hydrophobic condition and are then more readily wetted by organic media.

Some organic pigments are coated with resin to facilitate dispersion; these usually carry special suffixes, e.g. ED–P [31] (easily dispersed pigment). The pigment aggregates consist of an open lattice formed by a small quantity of resin. This is readily penetrated by paint media and dissolved, leaving the pigment particles in a dispersed condition. Such pigments can be dispersed readily by a high-speed stirrer.

Surface treatment is also applied to certain types of pigment or extender to check or prevent hard settlement, and an example is afforded by the deposition of stearic acid on whiting or on precipitated calcium carbonate. The adsorbed acid increases the volume of the particles considerably and leads to an appreciably lower overall specific gravity. In paints these materials are usually mixed with heavier pigments so as to check the rate of settlement and, in addition, the presence of the bulky particles in the sediment prevents the heavier pigments from packing into a hard cake. The sediment remains soft and is readily reincorporated on stirring.

The quantity of surface-coated calcium carbonate employed in a paint formulation should not exceed 10 percent by weight of the total pigment. Larger amounts are likely to interfere with the flow properties of the paint.

Certain grades of titanium dioxide, mainly the rutile type, are surface treated with oxides of zinc, aluminium, silicon and zirconium to improve their dispersibility and weathering properties (see page 54). Organic compounds also are incorporated.

Fastness to light

The ability to retain its colour when exposed to light is an essential property of a pigment. The wavelengths responsible for the photochemical breakdown lie in the ultraviolet waveband.

In addition to the effect on certain pigments, ultraviolet radiation can cause colour changes, most commonly yellowing or darkening, in some types of oxidizing media. When assessing lightfastness properties of a pigment it is essential, therefore, to ensure that the organic binder is unaffected, and for this reason acrylic media are commonly employed.

The lightfastness of a pigment depends on a number of factors, the most important of which are chemical constitution, purity, physical condition, and conditions of exposure.

Chemical constitution. Most inorganic pigments, in the pure state, are fast to light and, with a few exceptions, are fast in admixture with other pigments. Organic pigments, however, show a wide variation in lightfastness both in the pure form and when mixed with white ("reduced"). It is an interesting fact that the fastness to light (and the general stability) of organic pigments improves with increasing molecular size and complexity.

Purity. The term "purity" in this instance means freedom from admixture with other pigments. A pigment may be fast to light in the pure condition but may lose this fastness when diluted with white (or other pigment). The extent of the loss depends on the degree of dilution, and consequently lightfastness figures are determined at internationally agreed ratios of coloured pigment to white (usually titanium dioxide). They constitute the ISO (International Standards Organization) Standard Depths and are described in BS 2661 [32]. Ratios commonly employed vary between 1/10 and 1/200.

Physical condition. Lightfastness is little affected by particle size, but crystal structure and lightfastness are sometimes related, as in the lead chromes. Again, although it is not an example of fading, the anatase form of titanium dioxide is affected by light to a greater degree than the rutile form.

Conditions of exposure. The lightfastness of a pigment is important only when the pigment is dispersed in a medium, and the nature of this medium can affect the performance of the pigment. When lightfastness alone is measured (described below) the film of pigment dispersion is exposed to ultraviolet radiation under controlled conditions. In practice, however, films may be exposed to the combined effects of sun and rain. The latter penetrates the film to a greater or lesser degree and can affect the performance of the pigment. The term *durability* is generally used to denote the performance of the paint film on exterior exposure.

Measurement of lightfastness

The basis of the method is the assessment of the degree of fading which takes place when films of the pigment, and the standard, in the same medium are exposed simultaneously to the same conditions. This can be sunlight but is more usually the radiation from a carbon arc lamp housed in a Fugitometer (Fig. 2.13).

For many years Madder Lake was employed as a standard for lightfastness and was assigned an arbitrary value of 10. The degree of fading of other pigments was then assessed visually against this standard and they were assigned values of 1 (very poor) to 10 (good).

A more scientific method is now employed whereby a closer and more accurate assessment can be made. The method is described in BS 1006 [33] and involves the use of a range of eight blue wool specimens which have

Courtesy: J. B. Marr

Fig. 2.13   Fugitometer

been prepared with dyes of differing lightfastness. Thus, No. 1 has very low, and No. 8 very high, lightfastness. Each standard is approximately twice as fast to light as the preceding one.

The pigment is dispersed in a suitable medium and the dispersion applied to a card and allowed to dry. This is exposed either to daylight under glass, or to a carbon-arc lamp, at the same time as the series of blue wool specimens. At the end of the exposure period the sample is compared with the standards, and the degree of fastness quoted is the number of the wool specimen which has faded to the same degree. When changes in colour are involved, a grey wool scale is employed. This is described in BS 2662 [34].

In the fugitometer the specimens are exposed round the periphery of the drum and arranged so that part of the specimen is shielded from direct exposure. A carbon-arc rated at approximately 1500 to 1800 watts provides the ultraviolet light. Temperature is controlled by means of a fan in the base of the instrument.

Carbon-arc lamps produce radiation which is much richer in the active ultraviolet wavelengths than normal daylight, and results can sometimes be misleading. The spectral energy distribution of the radiation from a xenon-arc lamp is closer to that of normal daylight and these lamps are replacing the carbon-arc type.

Lightfastness figures can be obtained from the pigment manufacturers' literature, but a simple comparison of a pigment with one of known lightfastness (and the same type) may be needed. In this case the expensive blue wool patterns are not required, and use can be made of the method described in BS 3483, Part A3 [18] which also involves the use of the fugitometer.

Resistance to heat

Many paint films encounter heat either during a stoving process, which may be in the region of 120°C to 200°C (248° to 392°F) for periods of a few minutes to an hour, or they may be applied to surfaces which remain at high temperatures for long periods.

It is essential that the pigments used in such paints should be resistant to the temperatures encountered and they have, therefore, to be chosen with care. Pigments vary considerably in their resistance to heat and this property depends largely on the chemical composition. Most inorganic pigments—and many organics—will withstand 120°C (248°F) for short periods without decomposition. Above this temperature, however, the number of resistant pigments decreases progressively. An indication of heat resistance is given under the pigment classes. A method for the comparison of the heat stability of pigments is described in BS 3483, Part C8 [18].

Solubility ("bleeding") of pigments

"Bleeding" is the term used to describe the discoloration or staining which sometimes occurs when white or light coloured paint is applied over a deeper colour. It is generally (but not invariably) due to the solubilizing effect of solvents in the second coat on pigments in the first coat. The organic Para-nitraniline Red is well known for this tendency. A pigment which is very slightly soluble in the paint medium can function quite satisfactorily, provided the paint is not overcoated as mentioned above.

Solubility in water

With one or two exceptions, notably strontium and zinc chromes, inorganic pigments are insoluble in water. Among organic pigments the majority are insoluble in cold water, but a few tend to bleed in hot water.

Solubility in solvents, resins and oils

Inorganic pigments are insoluble. The solubility of organic pigments varies with the pigment type as well as with the nature and degree of substitution. Generally toners and lakes are less soluble than the pigment dyestuffs, but strong solvents of the ketone and ester types give greater bleeding than the more widely used hydrocarbon solvents. Very few pigments result in bleeding with the latter.

A method for the comparison of pigments for their resistance to bleeding is described in BS 3483, Part C7 [18].

Water-soluble matter in pigments

In the manufacture of many pigments, water-soluble chlorides and sulphates are produced as by-products. These are sometimes difficult to remove completely and for many uses their presence is not detrimental. If, however, the pigments are used in primers for metals, traces of these salts can actually stimulate corrosion, and for this reason it is advisable to determine the soluble salt content of the pigment before use. This is best carried out by the method described in BS 3483, Parts C2 and C6 [18].

TOXICITY OF PIGMENTS

The pigments which can exert pronounced toxic effects in the human body are largely confined to certain compounds of antimony, barium, cadmium, chromium, lead and mercury. (Mercury compounds are used to a limited extent as fungicides in water-based paints and as toxins in anti-fouling compositions for ships, but their use is being discontinued.) In order that a material shall be an active poison it must be soluble in the juices of the human body, and an insoluble material, although it may be a compound of a metal whose soluble salts are poisonous, will pass through the body without causing injury. For example, the soluble salts of barium, as well as the carbonate, are active poisons, but barium sulphate is insoluble and is, in fact, administered as a barium meal before X-ray photographs of the alimentary tract are taken. The most common of the toxic pigments are those containing lead, which is a cumulative poison; adsorption of small amounts over a period can result in chronic poisoning. For this reason the use of lead pigments is declining, the pigments having been eliminated from certain types of paint.

The Lead Paints (Protection against Poisoning) Act, 1926, with subsequent amendments and additions, was designed to protect operatives who come into contact with lead pigments or lead paints. The pigments can enter the body through the respiratory or digestive systems during the handling of the pigments themselves or during the application (particularly spraying) of paints containing them. Protection during the handling of the pigments can be provided by suitable clothing, respirators, and efficient fume extractors.

For lead paints to be applied by spraying, the Act lays down a maximum figure (5 percent) for the lead compounds soluble in 0·25 percent hydrochloric acid, which is the concentration in normal digestive juices. This soluble-lead figure is calculated as lead monoxide (PbO) and is expressed as a percentage on the pigment. It is determined by the method described in BS 282:1963 [35].

Under the conditions of this method white lead and litharge are completely soluble, but the soluble lead contents of lead chromes and chrome greens vary according to the composition. Figures for the soluble lead content of these pigments are usually available from the suppliers, but if they are used in admixture with other pigments it is necessary to determine the figure for the mixture.

Another possible source of poisoning from pigments is the paint applied to children's toys, and limits to the content of certain metals are specified in the Toys (Safety) Regulations, 1974 [36]. The maximum content of "total" lead is 2500 parts per million of the dry paint film; limits are also applied to "soluble" antimony, barium, cadmium and chromium.

Paints for decorative purposes contain less than one percent of lead to minimize the risk to small children, but where lead pigments continue to be used for certain types of industrial maintenance work, containers must bear the legend "This paint contains lead" if the total lead content exceeds one percent.

Factory regulations

The Health and Safety at Work Act, 1974, is concerned, *inter alia*, with the purity or degree of contamination of the atmosphere in places of work, and strict controls are in force.

Threshold Limit Values (TLV) have been assigned to most paint raw materials. These figures represent concentrations of air-borne substances below which the majority of workers can be exposed repeatedly without ill effect. They are time-weighted average concentrations which allow brief exposure to higher concentrations, provided it is followed by exposure to equivalent lower concentrations. "Ceiling values" are assigned to certain materials; these must not be exceeded.

Threshold Limit Values are published by the American Conference of Governmental Industrial Hygienists (ACGIM) who produce an annual list. Figures are sometimes amended in the light of experience.

TLVs are expressed either as parts per million (of air) or as milligrams per cubic metre. Some solids, generally in the extender class, are classed as "nuisance particulates", and their TLVs are expressed as millions of particles per cubic foot or per cubic metre.

Toxicity ratings

Toxic materials, in general, are given a "toxicity rating" designated LD_{50} which is the dose which will kill 50 percent of the particular class of animal to which it is administered. It is expressed as milligrams or grams per kilogram body weight. Paint components can be classified roughly as follows [37].

LD_{50} value	Toxicity
1 mg (or less)	High
100 mg	Moderate
1 gram	Slight
15 grams (or more)	Relatively innocuous

In the following chapters the degree of toxicity and TLV (when appropriate) are indicated under the individual materials.

The majority of organic pigments are considered to be reasonably safe. One or two, e.g. chromium compounds of certain dyestuffs, can produce slight skin irritation.

References to Chapter 2

[1] ISHIHARA, S., *Tests for Colour Blindness*. Lewis, London, 1959.
[2] Colour Matching Aptitude Test. Federation of Societies for Coatings Technology, Philadelphia, U.S.A.
[3] CLULOW, F. W., *Colour: its Principles & their Application*, p. 96. Fountain Press, London, 1972.
[4] Manufacturers' Engineering & Equipment Corporation, Pennsylvania, U.S.A.
[5] Instrument Development Laboratories (Kollmorgen Corpn), Massachusetts. CLULOW, F. W., *ibid*.
[6] Paint Research Association, Teddington, Middx. See also ISHAK, I. G. H., The fibre optics colorimeter and its applications in paint manufacture. *JOCCA*, 1971, **54**, 129.
[7] The Tintometer Ltd, Salisbury, Wilts.
[8] CLULOW, F. W., *ibid*., p. 106.
[9] Diano Corporation, Foxboro, Massachusetts.
[10] Pye-Unicam Ltd, Cambridge.
[11] Beckman-RIIC Ltd, High Wycombe, Bucks.
[12] BOC Automation, Daventry, Northants.
[13] KUBELKA, P. & MUNK, F., *Zeit. Tech. Physik*, 1931, **12**, 593.
[14] See also WRIGHT, W. D. *The Measurement of Colour*. Hilger & Watts, London, 1969. CLULOW, F. W., *ibid*.
[15] British Standards Institution, 2 Park Street, London.
[16] British Colour Council (now disbanded).

[17] Society of Dyers & Colourists, Bradford, Yorks.
[18] BS 3483:1974. Methods of Testing Pigments for Paints. British Standards Institution.
[19] BS 3900:1965 onwards. Methods of test for paints. British Standards Institution.
[20] BS 3406:1961–63. Methods for the determination of the particle size of powders. British Standards Institution.
[21] ANON., *J. Appl. Chem. Abs.*, 1962, i–612; 1963, ii–238.
[22] BOSTOCK, W. A., A sedimentation balance for particle size analysis in the sub-sieve range. *J. Sci. Instr.*, 1952, **29**, 209.
[23] BERESFORD, J., Size analysis of organic pigments using the ICI–Joyce Loebl disc centrifuge. *JOCCA*, 1967, **50**, 594.
[24] COSLETT, V. E., *Practical Electron Microscopy*. Butterworth, London, 1951.
[25] HALL, C. E., *Introduction to Electron Microscopy*. McGraw-Hill, New York, 1953.
[26] HERDAN, G., *Small Particle Statistics*. Butterworth, London, 1960.
[27] Coulter Electronics Ltd, Harpenden, Herts.
[28] BRUNAUER, S., EMMETT, P. H. & TELLER, E. J., *J. Am. Chem. Soc.*, 1938, **60**, 309. See also SING, K. S. W., The adsorption of gases and the characterisation of the surface properties of pigments. *JOCCA*, 1971, **54**, 731.
[29] BERESFORD, J., CARR, W. & LOMBARD, G. A., Surface areas of pigments. *JOCCA*, 1965, **48**, 293.
[30] CARMEN, P. C. & MALHERBE, P. le R., *J. Appl. Chem.*, 1951, **1**, 105.
[31] ICI Dyestuffs Division.
[32] BS 2661:1961. Methods for determination of colour fastness of textiles. British Standards Institution, London.
[33] BS 1006:1971. Method for the determination of fastness to daylight of coloured textiles.
[34] BS 2662:1961. Grey scale for assessing change in colour.
[35] BS 282:1963. Lead chromes and zinc chromes for paints.
[36] The Toys (Safety) Regulations, 1974. H.M. Stationery Office, London. See also BS 3443:1968. Toy Paints: Code of Safety Requirements for Children's Toys and Playthings. British Standards Institution.
[37] BRUNNER, H., in HESS, M., *Paint Film Defects*, ch. 8, 3rd edn. Chapman & Hall Ltd, London, 1979.

3   The processing of pigments

The maintenance of high quality in paints is possible only if the ingredients themselves are of high quality. Thus the pigments used should possess as narrow a range of particle size as possible and should be free from coarse particles, ultra-fines, and harmful impurities.

It is convenient, for the present discussion, to divide pigments and extenders into two classes: (i) those derived from natural sources, and (ii) those produced by chemical reactions. The processing of materials in class (i) is, broadly speaking, size reduction followed by grading by levigation or air-flotation. A large proportion of class (ii) materials is produced by precipitation which, in some cases, is followed by calcination. Other processes involve calcination or vapour-phase oxidation.

MINERAL PIGMENTS AND EXTENDERS

These materials are either mined or quarried and are obtained in the form of coarse lumps. The lumps are broken down to a coarse powder which is subjected to further grinding in one of the mills described below.

GRINDING

Edge runners

This type of mill consists of a circular, flat-bottomed cast-iron trough in which one or two heavy chilled-iron (hardened) rollers rotate on axes which are geared to a central shaft. In this way the rollers traverse the trough and produce a grinding effect which is a combination of crushing and shear. "Ploughs" are attached to the roller housing to deflect the materials under the rollers. Discharge is by a handwheel-operated slide in the bottom of the trough.

Edge runners used for pigment grinding usually have two rollers and are fitted with covers as a safeguard against dust and fire. These mills are very slow and have been largely superseded by more efficient machines.

Ball mills

Ball mills used for pigment grinding are similar in construction to those used in paint production and consist of a horizontal steel cylinder which is rotated on its axis. The mill contains a charge of steel balls, and the grinding action arises from the cascading of these balls when the mill is rotated (Fig. 3.1). For each size of mill there is a critical speed which ensures optimum cascading. Very low speeds give little cascading, whilst high speeds will result in the balls and charge being carried round by centrifugal action.

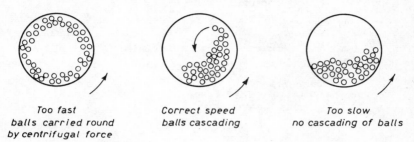

Too fast
balls carried round
by centrifugal force

Correct speed
balls cascading

Too slow
no cascading of balls

Fig. 3.1   Action of ball mills

The gross volume occupied by the balls should be about 40 to 45 percent of the total volume of the mill, and for maximum efficiency the material to be ground should just fill the voids in the ball charge.

These mills are used for both dry and wet grinding and for the former purpose are often fitted with an exterior dust casing. They are batch mills and in dry grinding suffer considerable wear on the balls and steel lining. When white pigments are ground the mill is lined with silica or porcelain blocks, and the steel balls are replaced by steatite or high-density alumina balls. However, the lower specific gravity of these materials results in much longer grinding times than when steel balls are used.

The micronizer

Size reduction in this mill takes place as a result of inter-particle attrition. The mill contains no moving parts and the energy required is derived from high-velocity compressed fluids entering the grinding chamber through tangential jets. For this reason it is classed as a fluid energy mill.

The mill, shown in Fig. 3.2, consists of a flat cylindrical grinding chamber fitted with inlet and outlet nozzles. The former are arranged tangentially round the outside and the compressed fluid — compressed air or superheated steam — creates a violent turbulence in the mill. The material to be ground is introduced by a venturi through which the compressed air or superheated steam is passed.

The violent turbulence and high velocities result in rapid size reduction by attrition. The fine particles are taken off in a stream of exhaust fluid and enter a collector. Uniformity of particle size is controlled by regulating the rate of feed and/or the pressure of the grinding fluid. The micronizer is used

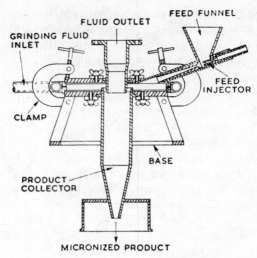

Courtesy: F. W. Berk

Fig. 3.2   The micronizer

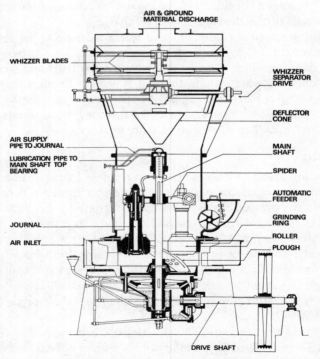

Courtesy: International Combustion Products

Fig. 3.3   Construction of the Raymond mill

mainly to produce powders with an average particle size of about five microns or less, but it can be used, if necessary, to produce coarser products.

Many natural pigments such as oxides of iron, as well as extenders such as barytes, talc, and dolomite are processed in these mills.

The Raymond mill

If a solid material is ground in an enclosed chamber and a vertical current of air is passed through, the finer particles are carried off in the airstream as they are formed. This principle, known as *air flotation*, is embodied in the Raymond mill (Fig. 3.3). The grinding chamber consists of a shallow pan, and a number of vertical rollers are suspended from a central shaft. When the latter rotates, the rollers press against the sides of the chamber as a result of centrifugal action. Coarse material is fed in and directed between the rollers and the casing. The finer particles are carried off in the airstream and a further separation is effected by passing the stream through rotating vanes in the upper part of the mill. These produce a cyclone effect, and the positions of the vanes determine the size of particle rejected. These fall back into the mill and are reground, whilst the finer particles are carried off in the airstream and collected.

The pin disc mill (Fig. 3.4)

This mill consists, basically, of two circular metal discs to which are attached steel pins arranged in concentric circles. The upper disc, with pins on the lower face, is fixed and has a central hole through which the pigment is introduced. The pins in the lower disc are on its upper face and the circles of pins pass between those of the upper disc. The lower disc is rotated, causing the material to be broken down between the pins, and centrifugal action throws the ground material off the edges of the discs where it is collected. Pin disc mills are suited to the softer types of pigment as they are pulverizing rather than grinding mills.

LEVIGATION

If a suspension of a pigment or extender in water is allowed to overflow from one to another of a series of vats of increasing diameter, the particles which settle out will be progressively smaller. The rate of flow of the liquid is reduced by the increasing diameter of the vats, so that the smaller particles are given the longer time required to settle out. In this way a material can be separated into a series of fractions of decreasing mean particle size, but there is no sharp line of demarcation between the fractions. Each will comprise a particle size range which will be determined by the size of the vats and the initial rate of flow of the liquid. There will be a degree of overlap with fractions above and below.

Courtesy: Kek

Fig. 3.4   Pin disc mill

This process, known as *levigation*, is employed to remove oversize particles and grit from several natural extenders where there is a plentiful supply of water near the source of the powder. It is generally preceded by a grinding operation. The deposit in each vat is filtered, dried, and finally reground.

MANUFACTURED PIGMENTS

The processes employed in the manufacture of pigments can be divided into the following types: (i) precipitation, (ii) precipitation followed by calcination, (iii) calcination, (iv) vapour-phase oxidation. A few pigments are prepared by methods not within this classification and these will be described under the pigments concerned.

PRECIPITATION

General principles

When two solutions, each containing a water-soluble salt, are mixed, a solid product will be precipitated if two of the radicals or ions present are capable of forming such a material. For example, if solutions of lead nitrate and sodium chromate are mixed, a yellow precipitate of lead chromate is produced, according to the equation:

$$Pb(NO_3)_2 + Na_2CrO_4 \rightarrow PbCrO_4 + 2\,NaNO_3$$

$$331 \qquad 162 \qquad\quad 323 \qquad 2 \times 85$$

The figures beneath the formulae are the molecular masses of the compounds involved. In practice it often happens that a product is improved if a slight excess of one of the reactants is used. In the above instance a slight excess of lead gives a brighter and more satisfactory product (Mid Chrome).

Another interesting example of precipitation is afforded by the manufacture of lithopone. The reaction in this case is somewhat unique in that both products of the reaction are insoluble. The process consists of mixing solutions of barium sulphide and zinc sulphate, when zinc sulphide and barium sulphate are precipitated:

$$BaS + ZnSO_4 \rightarrow ZnS + BaSO_4$$

Again, a slight excess of zinc produces a more stable product. The precipitated mixture is filtered, dried, and calcined.

Reduced pigments

If an insoluble base such as barytes is present when precipitation occurs, the precipitate will form on the particles of the base. Such pigments are known as *reduced*, and their tinting strength will vary with the amount of base present. Reduced pigments can also be made by dry-grinding the components.

Co-precipitated pigments

Some pigments contain more than one salt of a metal. For example, Lemon Chrome contains both lead chromate and lead sulphate. It is possible to make a pigment of this nature by mixing the two salts, but a far superior product is obtained if the two compounds are precipitated simultaneously. In this case a mixed solution of sodium chromate and sodium sulphate can be added to a solution of lead nitrate:

$$2\,Pb(NO_3)_2 + Na_2CrO_4 + Na_2SO_4 \rightarrow PbCrO_4\,PbSO_4 + 4\,NaNO_3$$

This produces a very intimate mixture of the two products and the process is known as co-precipitation. It is used to produce a number of pigments.

Conditions of precipitation

It is a well-known fact that the physical characteristics of a precipitate depend largely on conditions such as the concentration and temperature of the reacting solutions. This applies particularly to pigment manufacture, many of the properties of the pigment being influenced by the conditions of precipitation, of which the following are important.

Concentration of solutions

In general, fine precipitates are formed from dilute solutions, and coarser precipitates from more concentrated solutions. In practice, a compromise is often used. Whilst a fine particle size may be desirable for the end use, it can be difficult to filter. In a few cases the latter difficulty can be overcome by boiling the pigment suspension and so causing the particles to flocculate. Ten percent solutions have been found to be a very suitable compromise for a number of pigments, but there are several exceptions.

Temperature

High temperatures usually result in coarse precipitates, whilst finer precipitates are formed at lower temperatures. In most cases precipitation is carried out at 10°C to 20°C (50° to 68°F) unless the precipitate formed is too fine to filter. Prussian Blue is a pigment of this type and is therefore precipitated from boiling or near-boiling solutions.

Uniformity of stirring

The rate of stirring should be directly related to the rate at which the solutions are mixed. Considerable variation in particle size can result if these are not uniform.

pH value

For many pigments this is critical and the optimum value has, in most cases, been derived by experience. Variations from the critical value can affect colour and crystal form, as, for example, in the scarlet chromes (Chapter 5).

Precipitation plant

This is comparatively simple, the basic unit consisting of a precipitation vat holding up to 5000 gallons (22,700 litres), flanked by smaller dissolving vats.

The vats are constructed of oak or pitch pine and, when necessary, are lined with lead, rubber, or other acid-resisting material. In special cases, stainless-steel vats are used, but these are very expensive. Stirrers are of wood or coated metal and the vats are also fitted with open or closed steam pipes for heating, a cold water supply, and a bottom outlet to the filtration plant.

FILTRATION

The filtration of the pigment suspension from the precipitation vat can be carried out by either a filter press or a vacuum filter.

Filter presses

In these the suspension is pumped into chambers lined with filter cloths. The aqueous solution passes through the cloths to waste, and the pigment, in the form of a paste, remains in the chambers. Clean water is then passed through the paste to remove soluble by-products and the paste is "dewatered" by blowing air through the press. Some water remains and the product is then ready for drying or flushing.

Vacuum filters

Vacuum filters are of two general types which can be designated as the *rotary vacuum filter* and the *leaf type filter*.

Rotary vacuum filter

The general principle of the machine is illustrated by Fig. 3.5. A large drum is rotated on a horizontal axis and is partly submerged in a trough of the pigment slurry. The ends of the drum are solid, but the curved surface is constructed of perforated plate to which filter cloth is attached. Vacuum is applied to the interior of the drum. As the latter rotates, the water is drawn into it and the pigment collects as a paste on the outside, from which it is removed by a scraper. In some cases the drum is designed so that the adherent pigment is washed and partially dried by air-blowing before reaching the scraper.

It is necessary for the level of the pigment slurry in the trough to be kept constant, and the rotary vacuum filter is therefore admirably suited to continuous process filtration.
(Chapter 5).

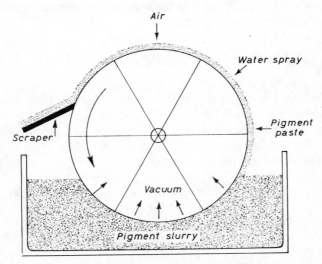

Fig. 3.5   Principle of the rotary vacuum filter

Leaf type vacuum filter

In this type of filter rectangular frames are made up of a number of leaves and attached to a vacuum line. The leaves are then covered with filter cloths. When the frames are lowered into a pigment slurry and vacuum applied, the liquid passes through the cloths and a layer of pigment paste builds up on the outside. When the layers are sufficiently thick the frames are lifted from the slurry and lowered into tanks of water for washing and the filtration repeated. After washing, the pigment cakes are removed and transferred to the drying ovens. This type of filter can be used for batch or continuous production.

DRYING

The conditions of drying can influence the texture and colour of many pigments. It is essential that the temperature be kept as low as possible otherwise hard clusters or *aggregates* of pigment particles can be formed and these can prove difficult to break down in subsequent operations. Ideally, the pigment when dry should be in the same physical condition as when filtered, but this is difficult to attain. Again, some organic pigments are very heat-sensitive when in the water-wet state and these must be dried at a low temperature.

The filter-press cakes are usually passed through an extruder in order to expose as large a surface as possible and spread in trays. These are either passed through a conveyor oven, where they meet a counter current of warm air, or placed in a low-temperature vacuum oven.

After drying, the pigment aggregates are reduced to their final size by grinding (for the harder aggregates) in mills of the type described earlier. For the softer aggregates, such as the organic and some inorganic pigments, the pin disc mill (page 45) is used.

FLUSHING [1]

Pigments can be divided into hydrophilic (water attracting) and hydrophobic (water repelling) types. If the wet filter-press cake of a hydrophobic pigment such as Hansa Yellow G (Chapter 8) is mixed with linseed oil or other organic medium in a suitable mixer, the pigment will pass into the organic medium and the water will separate out. This process is known as *flushing* and can be represented as follows:

$$\text{pigment/water} + \text{oil} \rightarrow \text{pigment/oil} + \text{water}$$

The water can be run off leaving the pigment dispersed in the medium. In addition to being used for hydrophobic pigments the process is also used with white lead. In this case the basic nature of the pigment assists the transfer to the organic medium.

Many of the inorganic pigments are hydrophilic and are therefore difficult to flush. The addition of suitable surface-active agents can convert them to the hydrophobic condition.

In spite of the obvious advantages in flushing, such as good pigment dispersion and avoidance of the costs of drying and subsequent dispersion, the process is used to a limited extent. This is probably due to the wide range of media and pigments used in modern paint manufacture.

CALCINATION

This term is applied to the process of roasting or heating to a high temperature in a suitable furnace. It is used as a process in itself or as further treatment for materials which have been precipitated, such as titanium dioxide.

The rotary furnace is the most widely used type of calciner for bulk production. It consists of a long steel cylinder mounted on trunnions. It is inclined at a slight angle to the horizontal so that material introduced at the top rolls round slowly and is discharged at the bottom. Heating is provided by combustion of a mixture of gas and air injected at the lower end. This type of furnace is employed in the manufacture of titanium dioxide.

Other types of calciners include the muffle furnace, the reverberatory furnace, and earthenware or silica crucibles.

Subsequent processing

During the process of calcination a certain amount of sintering of the pigment often takes place, resulting in the formation of clinker. To assist in breaking this up, the hot mass is discharged into cold water, when the clinker disintegrates. The pigment is then filtered, washed, dried, ground and graded.

Examples of pigments produced by direct calcination are red lead and Ultramarine Blue. Lithopone and titanium dioxide are produced by precipitation followed by calcination. Among the natural pigments, calcination is employed to convert Raw Sienna and Umber to the Burnt varieties (Chapter 5).

VAPOUR-PHASE OXIDATION

The metals zinc and antimony can be volatilized if heated to a sufficiently high temperature, and the vapours combine with oxygen by admixture with air to produce pigments of fine particle size.

In the case of zinc, the volatilized metal is mixed with air and the non-volatile zinc oxide precipitates. The zinc oxide is carried by the airstream to a classifier. (If the metallic zinc is volatilized in absence of air and then cooled rapidly, zinc dust is produced (Chapter 7).)

Antimony oxide is produced by vapour-phase oxidation of the metal, but the oxide is volatile at the temperature of reaction. It is removed to cooling chambers and cooled as rapidly as possible to produce a fine particle-size pigment.

There is also a vapour-phase oxidation method for the production of basic lead sulphate from galena.

The formation of carbon black (Chapter 7) by the controlled combustion of petroleum gases is also an example of this type of process.

Reference to Chapter 3

[1] GOMM, A. S., HULL, G. & MOILETT, J. L., Mechanism of the pigment flushing process. *JOCCA*, 1968, **51**, 143.

4 White pigments and extenders

TITANIUM DIOXIDE (CI Pigment White 6)

Sources

The two principal sources of titanium are the ores ilmenite and rutile. Ilmenite is a black ferrous titanate which occurs in several parts of the world, the most important being Norway, Travancore (India), United States, Canada, and Sweden. Rutile is a natural titanium dioxide which also occurs in a number of places and has been found to be a very suitable raw material for the preparation of the pigment by the chloride process (*below*).

Compounds of titanium are very widely distributed among a number of minerals, but in small quantities. Titanium is also present in many plants and can be detected in the residue when these are ashed.

MANUFACTURE OF TITANIUM DIOXIDE

The sulphate process

This uses ilmenite as the raw material. The ore is crushed, separated from associated materials, ground and dried. It is then dissolved in oleum. This reaction is smooth, whereas with concentrated sulphuric acid a violent reaction takes place. The solution in oleum is diluted gradually, first with sulphuric acid and then with water. This yields a solution of the sulphates of titanium and iron (ferrous and ferric) together with insoluble siliceous matter. The latter is filtered off. Scrap iron is then added to the solution to reduce all iron to the ferrous condition. The solution is separated, filtered, and transferred to a vacuum evaporator where most of the ferrous sulphate crystallizes out.

The next operation is a crucial one since it controls both the particle size and the nature of the resulting pigment. The solution is boiled to hydrolyse the titanium sulphate to titanium hydroxide, and the conditions of hydrolysis must be rigidly controlled. The precipitated hydroxide is washed thoroughly to remove acid and traces of iron, filtered, and then calcined in a rotary furnace at about 800°C (1472°F).

Finally the titanium dioxide is given the appropriate surface treatment. The sequence of operations is as follows:

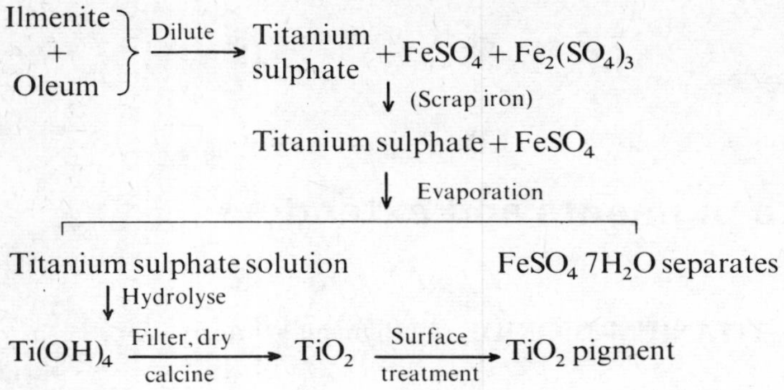

$$\left.\begin{array}{l}\text{Ilmenite} \\ + \\ \text{Oleum}\end{array}\right\} \xrightarrow{\text{Dilute}} \begin{array}{l}\text{Titanium} \\ \text{sulphate}\end{array} + FeSO_4 + Fe_2(SO_4)_3$$

$$\downarrow \text{(Scrap iron)}$$

$$\text{Titanium sulphate} + FeSO_4$$

$$\downarrow \text{Evaporation}$$

Titanium sulphate solution $FeSO_4\ 7H_2O$ separates

$$\downarrow \text{Hydrolyse}$$

$$Ti(OH)_4 \xrightarrow[\text{calcine}]{\text{Filter, dry}} TiO_2 \xrightarrow[\text{treatment}]{\text{Surface}} TiO_2 \text{ pigment}$$

The process outlined above yields the anatase type. The rutile type is produced by "seeding" with rutile particles or addition of a zinc salt before the hydrolysis stage.

The chloride process

This is of more recent introduction and involves passing chlorine gas over a mixture of rutile ore and coke at about 900°C (1652°F). The rutile ore contains 95–96 percent TiO_2 and the following reaction takes place:

$$2TiO_2 + 3C + 4Cl_2 \rightarrow 2TiCl_4 + 2CO + CO_2$$

The titanium tetrachloride volatilizes but is mixed with volatilized ferric chloride formed from iron impurity in the ore. On cooling the gases, the ferric chloride condenses at a temperature above the boiling point of the titanium tetrachloride. The latter is then condensed to a liquid by further cooling and is purified by distillation.

The titanium tetrachloride is oxidized to titanium dioxide by burning in a stream of oxygen at about 1000°C (1832°F):

$$TiCl_4 + O_2 \rightarrow TiO_2 + 2Cl_2$$

The chlorine produced is recirculated. To ensure complete formation of the rutile type, the oxygen contains a trace of moisture; in addition, a small amount of aluminium chloride is mixed with the titanium tetrachloride and oxidized at the same time.

After washing thoroughly to remove traces of acid, the pigment is given the required surface treatment and dried.

The ease with which the titanium tetrachloride can be purified, coupled with the close control of the oxidation process, results in a pigment of outstanding whiteness and constant particle size range.

Surface treatment

It has been found that the general working properties (e.g. wettability, ease of dispersion) and some aspects of the performance (resistance to chalk-ing) of titanium dioxide can be improved by surface treatment with oxides or hydroxides of aluminium, silica, and zinc. These are precipitated on to the pigment surface from solutions of the respective salts, followed by filtration, washing and drying. Surface treatment reduces the catalytic activity of the pigment surface, that of rutile being reduced to a very low value.

The relative proportions of the oxides and the amount of coating depend on the purpose for which the particular grade of titanium dioxide is intended, e.g. decorative gloss finishes, industrial stoving enamels, emulsion paints, etc. The TiO_2 content of the treated pigment can vary between 90 and 95 percent.

Properties of titanium dioxide

Titanium dioxide (Fig. 4.1) has greater opacity and tint resistance than any other white pigment and consequently has largely replaced other white pigments in a number of applications. It is non-toxic, resistant to heat and to the media and solvents used in paint manufacture. It also possesses great chemical stability and is dissolved only by hot concentrated sulphuric acid — and this process is slow.

The pigment exists in two forms known as anatase and rutile. Both forms fall into the tetragonal system, but anatase has the lower density arising from a greater distance between the titanium and oxygen atoms. When exposed to ultraviolet radiation both forms absorb strongly. Anatase becomes highly "excited" and exhibits intense surface catalytic activity.

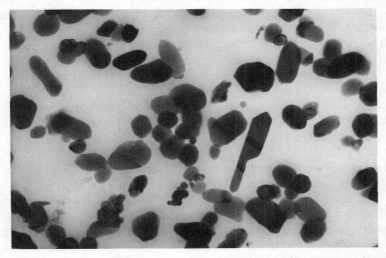

Courtesy: Paint Research Association

Fig. 4.1   Photomicrograph of titanium dioxide (× 30 000)

Excitation and catalytic activation of rutile is very much less, most of the absorbed energy being dissipated as heat (see Chalking, *below*).

The ultraviolet absorption band of rutile overlaps the far blue end of the visible spectrum (up to about 410 nm) with the result that in daylight the reflected light is deficient in blue and gives rutile the characteristic yellowish cast.

The physical characteristics of the two grades are as follows:

	Rutile	*Anatase*
Colour	Slight yellowish cast (chloride types are cleaner and brighter than sulphate process types)	Cold bluish-white
Specific gravity	4·0–4·1	3·7–3·8
Refractive index	2·71	2·55
Oil absorption	17–24	19–22
Relative tint resistance	1·2–1·4	1·0
Resistance to "chalking"	Good	Poor
Mean particle size range	0·2–0·3 μm	0·18–0·25 μm

Chalking

This is a condition of a paint surface when free or "unbound" pigment is present as a result of decomposition of the surface layer of binder. Gentle rubbing of the finger-tip across the surface removes the free pigment as a white powder or "chalk".

Paint films pigmented with both rutile and anatase types of titanium dioxide will "chalk" after periods of exposure to weather. With anatase pigmentation the onset of chalking is much more rapid, and the photochemical breakdown of the film more severe than with rutile. In fact the rutile grades, particularly the surface-treated types commonly used, cause only a slight tendency to chalking of the film after exposure for a number of years. The rutile grades are in general use in exterior finishes, but it should be mentioned that the nature of the binder can influence the chalking tendency of a paint film.

Occasionally paint films are designed to give a "controlled" rate of chalking in order to shed dirt and to maintain a clean white appearance. For this purpose mixtures of rutile and anatase grades of titanium dioxide are employed.

The breakdown of the surface layer of binder which leads to chalking is due to accelerated oxidation in the presence of titanium dioxide. The latter acts as a catalyst, the anatase grade being much more active than rutile. The fact that titanium dioxide will catalyse certain oxidation reactions has long been known, and Murley [1] as well as Kennedy et al. [2] have demonstrated that, under ultraviolet radiation, anatase absorbs oxygen to a greater degree than does rutile. In the paint film both oxygen and certain components of the binder are thought to be adsorbed on the activated titanium dioxide surface. Adsorption of reactants at the surface of a solid generally results in a lowering of the activation energy of the reaction (in this case oxidation) which then proceeds more rapidly. Myers [3] considers that when titanium dioxide is converted to an excited state on exposure to ultraviolet radiation, electron transfer reactions take place with oxygen and water to produce oxidizing agents.

Although the oxidation and subsequent decomposition of the binder may commence at the titanium dioxide surface, it occurs throughout the surface layer of binder, much of which is not in actual contact with pigment. It would appear that a certain polymer species becomes oxidized at the titanium dioxide surface and either diffuses away and catalyses the oxidation of other molecules [1] or possibly sets up a chain reaction by a free radical or activation mechanism.

Uses of titanium dioxide

It is the most widely used white pigment. The two grades, with suitable surface treatment, are used in the following types of paint:

Rutile: decorative, maintenance and industrial gloss finishes; decorative semi-gloss, eggshell and matt finishes; marine paints; undercoats; emulsion paints for interior and exterior use.

Anatase: used in industrial finishes where good whiteness is important but where weather resistance is not required, e.g. domestic and hospital equipment.

Identification of titanium dioxide

The outstanding chemical resistance of the pigment enables it to be separated from most other pigments (but not from some extenders which, however, do not interfere with the tests). Titanium dioxide may be recognized by the following simple test.

A small quantity of the pigment is placed in a depression on a spotting tile and moistened with concentrated sulphuric acid. Addition of a few drops of hydrogen peroxide solution results in an orange or orange/yellow coloration.

Toxicity of titanium dioxide

The pigment is regarded as non-toxic, but dust arising during handling is classed as a "nuisance particulate".

$$\text{TLV (TiO}_2 \text{ dust)} = 10 \text{ mg/m}^3$$

Spindrift [4] is an interesting product which utilizes the greater scattering power of titanium dioxide when in contact with air compared with that when in contact with an organic medium. The greater scattering power results in improved opacity. In Spindrift, particles of titanium dioxide, surrounded by air, are enclosed in spherical shells of a polymer — usually a co-polymer of styrene and an unsaturated polyester. The spheres range in diameter from 3 to 50 micrometres. The use of Spindrift is confined to semi-matt and matt finishes; it has been found to function successfully in emulsion paints.

ZINC OXIDE (CI Pigment White 4)

This pigment has been used in paints for very many years and is known to artists as *Chinese White*.

With the advent of titanium dioxide and the use of alkyd and other synthetic resins the use of zinc oxide as a pigment in paints has declined very considerably, and today little is used for this purpose in this country. A certain amount of colloidal zinc oxide (*below*) is thought to assist gloss retention in some enamels and is used for this purpose.

Manufacture of zinc oxide

This is a vapour-phase oxidation and is based on the fact that metallic zinc begins to volatilize when heated above its melting point (419°C, 786°F), and the vapour combines rapidly with oxygen to form the non-volatile zinc oxide. Two methods are in use, known as the direct and indirect processes.

Direct process

In this process the oxide is produced directly from the ore in one operation. The ore — preferably a carbonate — is mixed with coal or coke and heated in a furnace. The ore is reduced to metal which volatilizes and is then mixed with a current of air. Zinc oxide is formed and is carried by the stream of air into suitable collecting chambers.

The product obtained by this process contains impurities, the chief of which is basic lead sulphate (up to 5 percent).

Indirect process

Metallic zinc, known industrially as *spelter*, is first produced and in this operation ores such as the sulphide (zinc blende) can be used. The spelter is heated in crucibles to volatilize the metal, which passes out through holes in the lids. The metal vapour is then mixed with a stream of air and the zinc oxide produced is carried into the collecting chamber. This oxide usually contains less than 0·2 percent of impurities.

Properties of zinc oxide

The bulk of the pigment produced is the amorphous type, but by adjustment of the manufacturing conditions it is possible to produce a crystalline ("acicular") or a very fine ("colloidal") type. Amorphous zinc oxide (indirect type) possesses the following characteristics:

Specific gravity 5·65   Oil absorption 16–23.
Refractive index 2·0   Mean particle size range 0·2–0·35 μm

Zinc oxide is a reactive pigment and forms zinc compounds or "soaps" when mixed with drying oils or varnishes, the extent of reaction depending on the acid value and the degree of polymerization of the medium. Films deposited from zinc oxide dispersed in such media exhibit a high gloss but undergo progressive embrittlement. Such paints were formerly used for interior decorative work. The acicular grade, however, is less reactive and can be used in exterior finishes.

Uses of zinc oxide

In present-day paints the use is limited to certain specific rôles:

(i) It has pronounced fungistatic properties and is incorporated into emulsion paints for use in warm and damp climates where fungus attack is a major problem.
(ii) As mentioned earlier, a small quantity of the colloidal grade is sometimes included in certain alkyd-type decorative enamels to assist gloss retention. The amount employed is usually critical and depends on the nature of the medium.

Identification tests

Zinc oxide has the well-known property of turning yellow on heating and reverting to white on cooling. If it is moistened with cobalt chloride solution and re-heated a green coloration is produced.

The pigment is soluble in dilute mineral and acetic acids.

Toxicity of zinc oxide

The pigment does not present any significant hazard. "Fresh" zinc oxide fume, as produced in welding or flame-cutting of films containing metallic zinc, can cause "zinc fume fever" but the effects are temporary.

$$\text{TLV (ZnO)} = 5 \text{ mg/m}^3$$

ZINC PHOSPHATE (CI Pigment White 32)

A white, non-toxic pigment which is prepared by precipitation, it is assigned the formula $Zn_3(PO_4)_2\ 2H_2O$. It possesses the following characteristics:

Specific gravity 3·3	Oil absorption 18–20.
pH of aqueous extract 6·8–7·0	Mean particle size range 2–4 μm

It has poor opacity, but this is not a serious drawback as the main use of the pigment is in corrosion-resisting primers for steel and in undercoats. Zinc phosphate is compatible with and performs satisfactorily in a number of media including alkyds, chlorinated rubber, epoxy esters, two-pack epoxies, and polyurethanes.

The corrosion-resisting properties have attracted a number of investigators, but the mechanism of protection is not completely understood. According to Clay and Cox [5] zinc phosphate reacts by polarizing both anodic and cathodic areas as a result of slight solubility. On the other hand, the pigment appears to pack in the film in a manner which presents a high resistance to the passage of water molecules and salts. The protective qualities are likely to be a combination of electrochemical suppression of corrosion cells with a moisture barrier effect.

Identification test. Zinc phosphate is soluble on warming with 4M mineral acid. The solution gives the reactions of zinc and phosphate ions.

Toxicity. The pigment is regarded as free from toxic hazard.

ZINC PHOSPHO-OXIDE

A white non-toxic pigment marketed under the trade name Nalzin SC–1 [6] and which is claimed to possess excellent corrosion-inhibiting properties when used as a replacement for red lead. It is also claimed to be specially suited to electrodeposition. The physical characteristics are as follows:

Specific gravity 4·06   Mean particle size range $< 1 \mu$m
Oil absorption 50–70

A further interesting use of this pigment has been described by Davidson [7] who found it to be an excellent control agent for cedar stain.

LITHOPONE

This pigment which is sometimes known as *zinc white* has been used for many years and contains zinc sulphide and barium sulphate.

Manufacture

The basic reaction involved is double decomposition using solutions of barium sulphide and zinc sulphate. The reaction is somewhat unusual in that both products are insoluble in water:

$$BaS + ZnSO_4 \rightarrow ZnS + BaSO_4$$

The precipitate is a mixture of zinc sulphide and barium sulphate and of little interest as a pigment. It is filtered, washed, dried and then calcined at 700°C (1292°F) in an inert atmosphere to develop the pigmentary properties. The hot mass from the calciner is discharged into cold water to break up the *frit*, and then filtered, dried and ground.

Properties of lithopone
Specific gravity 4·2   Oil absorption 12–15
Refractive index 1·90   Mean particle size range 1–3 μm

It is a clean white and does not react with acid media, although it is attacked by mineral acids. Lithopone is very little used in solvent-based paints, but it is occasionally encountered in emulsion paints in which it can form part of the pigment mixture.

Toxicity. Lithopone is considered to be free from toxic hazard.

Zinc sulphide

This material is marketed under the proprietary name "Sachtolith" [8]. It is prepared by precipitation from a soluble zinc salt followed by similar treatment to lithopone. The pigment possesses very good opacity and tint resistance, but pigments containing zinc sulphide are generally considered unsuitable for use in paints for exterior exposure since zinc sulphide tends to oxidize to the water-soluble zinc sulphate which is leached out of the film. It is claimed, however, that certain grades of zinc sulphide are suitable for use in exterior paints [8].

ANTIMONY OXIDE (CI Pigment White 11)

Antimony oxide (Sb_2O_3) is also known under the proprietary name *Timonox* [9] and is a pigment whose usage in paint has declined in recent years. It is now used mainly for certain specialized purposes.

Manufacture

The chief source of the metal is the ore stibnite (Sb_2S_3) from which the metal is extracted by roasting with iron:

$$Sb_2S_3 + 3Fe \rightarrow 2Sb + 3FeS$$

Antimony oxide is manufactured from the metal and the process bears a certain resemblance to the indirect process for zinc oxide. The metal is volatilized by heat and the vaporized metal is oxidized to oxide in a current of air. Antimony oxide, however, differs from zinc oxide in that it is volatile at the temperature of formation and is carried by the airstream into condensing chambers where the solid pigment is deposited on cooling. The rate of cooling determines the particle size and crystalline form. Rapid cooling produces an amorphous pigment of fine particle size, whereas slow cooling gives a less desirable crystalline product. The oxidation reaction can be represented by the equation

$$4Sb + 3O_2 \rightarrow 2Sb_2O_3$$

Properties and uses of antimony oxide

Antimony oxide possesses the following physical characteristics:

Specific gravity	5·5
Oil absorption	11–13
Refractive index	2·20
Mean particle size range	0·5–2·0 μm

It possesses very good opacity and high stain resistance. It is non-reactive toward organic media and for this reason it was at one time used with zinc oxide to counteract the film embrittlement caused by the latter. It also works well in cellulose nitrate lacquers.

When exposed to sunlight, paints based on antimony oxide do not chalk and for some years the pigment was used in conjunction with anatase titanium dioxide to suppress chalking. With the introduction of the chalk-resistant rutile titanium dioxide the use of antimony oxide declined. The principal use of the pigment today is in fire-retardant paints where it is used in conjunction with a chlorinated medium.

Identification tests

On strong heating, antimony oxide melts and then volatilizes. The vapour will form needle-like crystals on cooling.

It is insoluble in dilute mineral acids but dissolves in concentrated hydrochloric acid. The solution gives an orange precipitate of antimony sulphide with hydrogen sulphide. If the solution in concentrated hydrochloric acid is diluted considerably with water, a white precipitate of oxychloride is produced:

$$SbCl_3 + H_2O \rightarrow SbOCl + 2HCl$$

WHITE LEAD (CI Pigment White 1)

This material has been known for many centuries and to artists was familiar as *flake white*. It consists of a basic carbonate of lead and the composition approximates to the formula $Pb(OH)_2 . 2PbCO_3$.

Manufacture

For very many years white lead was manufactured by very slow processes, the most famous of which was the stack process. A later and quicker variant was the chamber process used in this country and the Carter process used in the United States. All these methods were based on corrosion of metallic lead by steam, acetic acid and carbon dioxide, and they have been described in detail in earlier books. These processes are now virtually obsolete and most of the modern white lead is produced by precipitation processes.

The precipitation process

The basis of the process is the treatment of a solution of basic lead acetate with carbon dioxide under carefully regulated conditions. White lead is precipitated and normal lead acetate regenerated.

The basic lead acetate solution is produced by pumping a solution of normal lead acetate over a bed of spongy lead in the presence of air. The spongy lead is prepared by pouring the molten metal into water. Alternatively litharge can be used but is more expensive.

A circulatory system is employed and continued until sufficient lead has been taken up, as indicated by the specific gravity. The reaction can be represented thus:

$$Pb(CH_3COO)_2 + Pb + [O] + H_2O \rightarrow Pb(CH_3COO)_2 . Pb(OH)_2$$

The basic lead acetate solution is then pumped to a reaction vessel and treated with carbon dioxide under controlled conditions, when the white lead is precipitated:

$$3[Pb(CH_3COO)_2 . Pb(OH)_2] + 2CO_2 \rightarrow Pb(OH)_2 . 2PbCO_3 +$$
$$3Pb(CH_3COO)_2 + 2H_2O$$

After separation of the white lead by filtration the solution of normal lead acetate is re-circulated over the spongy lead.

The filtered white lead is then either dried off and sold as powder or converted to oil paste by flushing.

The precipitation process has several advantages over the earlier processes, one of which is the much shorter time required. From the user's point of view the chief virtue lies in the degree of control possible whereby the physical characteristics of the pigment can be controlled.

Properties of white lead

Specific gravity 6·7   Oil absorption 7·5–10·0

Refractive index 2·0

It is chemically reactive and forms lead "soaps" when mixed with linseed oil or varnish. These soaps are largely responsible for the elasticity of the films of white lead paints and for their protective qualities.

For many years white lead paint was the standard type of finish and was satisfactory under rural and marine conditions. In urban atmospheres the hydrogen sulphide present caused discoloration, the extent of which varied with locality.

The introduction of alkyd resins and rutile titanium dioxide which give durable gloss finishes with good flow has caused a decline in the use of white lead in finishing paints. Although the protective qualities of white lead oil paint were excellent, the flow and gloss retention were inferior to those of modern alkyds.

White lead is a highly toxic pigment and, as a lead compound, acts in the human system as a cumulative poison. It is completely soluble in the stomach acid (0·25 percent hydrochloric acid) and so is readily ingested.

Both the manufacture and use of white lead (in common with other lead pigments) are subject to the regulations of the Lead Paints (Protection against Poisoning) Act, 1926 and subsequent amendments.

On account of its high toxicity and the widespread use of titanium dioxide white lead has been phased out of the majority of paints, but it continues to be specified in some undercoats for certain types of exterior maintenance work.

Identification tests

On heating, white lead turns orange and finally yellow, due to the formation of litharge.

It is completely soluble in dilute acetic and nitric acids with evolution of carbon dioxide. Reaction takes place with hydrochloric acid, but the lead chloride produced is soluble only to a limited extent in the cold but dissolves on boiling. Treatment of the solution with hydrogen sulphide gives a black precipitate of lead sulphide. Addition of potassium iodide solution to the hot solution of lead chloride leads to the formation of golden-yellow spangles on cooling.

Treatment of white lead with dilute sulphuric acid results in an initial reaction which is rapidly stifled by the film of insoluble lead sulphate which forms on the surface.

BASIC LEAD SULPHATE (CI Pigment White 2)

The composition of this pigment approximates to the formula $PbO \cdot 2PbSO_4$, but variations occur due to fluctuations in quality of the raw material used.

Manufacture

Basic lead sulphate can be manufactured by roasting natural lead sulphide (galena) so that it volatilizes and is led into a second chamber where it is oxidized by a cold air blast. The pigment is then collected.

Precipitation methods are also employed and consist in treating an aqueous suspension of litharge or litharge mixed with finely divided lead with carefully regulated quantities of sulphuric acid.

Properties and uses

Basic lead sulphate is less reactive than white lead towards oils and varnishes and has not been used to the same extent in gloss paints and undercoats. It is toxic and the soluble-lead content is in excess of that permitted for use in spray paints. At the present time the main use is in certain types of primer for the underwater areas of ships' hulls. The pigment possesses the following physical characteristics:

Specific gravity	6·4
Oil absorption	12–15
Refractive index	1·93

Identification tests

Basic lead sulphate dissolves completely on boiling with medium strength hydrochloric acid, and the solution gives the usual qualitative tests for lead and sulphate.

It is, of course, subject to the Lead Paint Regulations.

EXTENDERS

These are, in general, chemically inert materials some of which are processed minerals while others are prepared chemically. They are characterized by the fact that their refractive indices fall within a narrow range and this is close to that of oils and resins. Consequently extenders are practically transparent in paint media and make no contribution to colour or opacity. They are opaque in water and can therefore be used as pigments in certain types of water paint.

When pure, extenders are colourless, but many natural products are coloured by impurities which are often difficult to remove.

The term *extender* suggests a material used to cheapen a product, and whilst it is true that on occasions extenders are used for this purpose they are more often used to confer desirable properties on a paint. For example, they can be used to control consistency, as flatting agents, and to prevent hard settlement of heavy pigment. The term "auxiliary pigment" might therefore be more appropriate.

Classification of extenders

Extenders are best classified by chemical composition, and the majority can be classified as compounds of the metals barium, calcium, aluminium and magnesium. The exception is silica, and the natural and prepared silicas are considered as a separate class.

BARIUM COMPOUNDS

These comprise the sulphate which occurs naturally as barytes or heavy spar and the precipitated form known as blanc fixe.

BARYTES

This mineral occurs extensively in many parts of the world, but the colour and quality vary considerably. Some of the purest grades are obtained in Germany and there are extensive deposits of high-grade material in the United States; other sources are Algeria and Greece. Home-produced barytes is generally of poor colour and the quality varies considerably.

In view of the variation in colour, barytes is classified as *white* and *off-colour*, the latter term covering a very wide range of materials.

Processing

If the mineral is a pure crystalline grade of barytes it is crushed, ground and graded by air-flotation. Off-colour grades are often treated with sulphuric or hydrochloric acid to improve the colour by removing impurities, which may be iron oxide or compounds of lead. Calcium sulphate or carbonate may also be present. As barytes is a heavy pigment and settles rapidly the acid can readily be washed out by decantation. The barytes is then filtered, dried, ground (much is now micronized) and graded by air-flotation.

Grades

Barytes is available in ordinary grades in which the particle size can range up to 20 to 25 μm, or in micronized form. In the latter it is possible to obtain a number of size ranges and these are marketed with an average particle size of 2 μm, 5 μm, 10 μm, etc. These micronized grades are now widely used and are displacing the ordinary grades. In addition they have a lower oil absorption than blanc fixe and for many purposes are displacing this extender.

Properties

Barytes possesses the following physical characteristics:

Specific gravity	4·5
Oil absorption	10–12
Refractive index	1·67
Mean particle size range	Varies according to grade

When pure it is practically transparent in organic media, but most commercial grades show some slight opacity due to impurities.

It is inert toward all paint media and chemically is very stable. It will dissolve only in concentrated sulphuric acid on heating but is reprecipitated on cooling or dilution.

Uses

Barytes is widely used in undercoats and primers in which it is thought to give physical reinforcement to the film. It is also used in fillers, stoppers, and some types of masonry finish.

Identification tests

Natural barytes is distinctly crystalline under the microscope, but the crystalline nature is less obvious in the micronized grades. This process tends to round off the corners and sharp edges.

In the platinum wire flame test, barytes gives the characteristic green-coloured flame of barium.

The material is extremely inert, and in order to confirm the presence of barium and sulphate it is necessary to fuse it with anhydrous sodium carbonate. Decomposition takes place according to the equation

$$BaSO_4 + Na_2CO_3 \rightarrow BaCO_3 + Na_2SO_4$$

The fused mass is extracted with water and filtered. The filtrate is acidified with dilute hydrochloric acid and barium chloride solution is added. A white precipitate indicates sulphate. The residue is dissolved in dilute hydrochloric acid, and then dilute sulphuric acid is added. Formation of a white precipitate indicates barium.

Toxicity. Barium sulphate is insoluble in the digestive fluids and is considered to be non-toxic. It is used in medicine as the basis of the "barium meal" given to patients before intestinal X-ray examination.

BLANC FIXE

This is a precipitated form of barium sulphate and generally contains about 98 percent $BaSO_4$. It can be produced by treatment of barium sulphide solution with a soluble sulphate such as sodium sulphate, according to the equation

$$BaS + Na_2SO_4 \rightarrow BaSO_4 + Na_2S$$

The product is washed well, filtered and dried.

An alternative method of production utilizes natural barium carbonate, witherite. This is dissolved in hydrochloric acid to produce barium chloride which is then treated with sulphuric acid:

$$BaCl_2 + H_2SO_4 \rightarrow BaSO_4 + 2HCl$$

This is followed by filtering, washing and drying.

Properties and uses

Blanc fixe possesses the following physical characteristics:

Specific gravity	4·3
Oil absorption	14–16
Refractive index	1·67
Mean particle size range	1–5 μm

As a precipitated product, blanc fixe has a finer texture and higher oil absorption than barytes. It dissolves more readily than barytes in concentrated sulphuric acid but otherwise shows the same chemical inertness.

It has been used as an extender in undercoats but has tended to be replaced by micronized barytes which gives better flow. In addition, the use of high percentages of blanc fixe in undercoats can lead to *sinkage* and loss of gloss in finishing coats.

Blanc fixe alone and as a co-precipitated base with alumina is fairly widely used as a base for organic pigments (Chapter 8).

In water paints, the finer texture of blanc fixe gives it greater opacity than ordinary barytes and confers smoother working properties on the paints.

Toxicity. See Barytes (*above*).

CALCIUM COMPOUNDS

The calcium-based extenders of interest in paint manufacture comprise the carbonate, as natural whiting or as the precipitated form; the double carbonate with magnesium as the mineral dolomite; and, to a lesser extent, the sulphate.

WHITING

The purer grades of this material are sometimes known as Paris White. Whiting is obtained when natural chalk (mainly $CaCO_3$) is reduced to a powder. Chalk consists of the remains of tiny creatures (coccoliths) laid down in ocean beds over very long periods of time. In electron micrographs of whiting, the ring-like structure of the coccoliths is often quite apparent.

Chalk comprises the basic structure of much of south-east and north-east England and of some hill ranges such as the Yorkshire Wolds and the Chilterns.

Processing of whiting

There are two general methods in use for the treatment of the chalk and the basic sequence of operations is the same, namely, grinding, drying and grading. The different processes arise from the fact that whitings differ in hardness and density. The softer types are treated by a wet process and the harder types by a dry process.

The wet process

The soft type of whiting is broken up by stirring in water. Any nodules of flint separate out, and the slurry is passed through hydrocyclones which separate the coarse material. The slurry is next placed in shallow tanks for settlement. The paste obtained after decanting off the upper liquor is filtered by press or rotary vacuum filter and dried. It is finally disintegrated and packed.

The dry process

The raw material is crushed and screened to remove the coarse particles and grit. It is then passed through a pin disc mill such as the Kek (Fig. 3.4) where it is reduced to a fine powder. Hot air can be blown through at the same time and serves both to dry the powder and to carry it to a cyclone for grading.

Alternatively the whiting can be given a preliminary drying and is then ground and graded in a roller mill of the Raymond type (Fig. 3.3).

Properties

In this country whiting is marketed in two forms, known as paint grades and putty grades. The former is obtained from the softer chalks of south-east England, whereas the harder putty grades are derived from the north-east of the country. Both grades contain about 96 to 97 percent $CaCO_3$, the remainder consisting of fine clay. The pinkish colour is due to traces of iron oxides.

The physical characteristics of the two grades are as follows:

	Paint grade	Putty grade
Specific gravity	2·5–2·8	2·5–2·8
Oil absorption	17–18	16–17
Refractive index	1·58	1·58
Mean particle-size range	2–5 μm	2–5 μm but with a number of particles over 10 μm

Whiting shows an alkaline reaction in water and is readily soluble in dilute mineral and acetic acids.

Uses of whiting

Glazing putty for wooden window frames consists of whiting dispersed in linseed oil. Whiting is used in many fillers and stoppers and, to varying extents, in certain undercoats and primers. The presence of a proportion of whiting in primers for steelwork is thought to be an advantage in providing an alkaline environment. Whiting is often incorporated in vinyl emulsion paints where it acts as a "buffer" against the possibility of a fall in pH value.

Identification tests

Whiting shows an alkaline reaction when mixed with distilled water. It is decomposed on strong heating leaving a residue of lime, and the platinum wire flame test gives the characteristic crimson colour of calcium. It is soluble in dilute mineral and acetic acids giving carbon dioxide and a solution which shows the usual reaction for calcium ions, i.e. it gives a white precipitate with ammonium oxalate under slightly alkaline conditions.

Coated whitings

The chemical reactivity of whiting enables it to react with some of the long-chain fatty acids such as stearic acid, whereby a coating of the stearate is formed over the surface. The amount of acid can be controlled up to 2 percent and the products then exhibit hydrophobic properties in contrast to the hydrophilic properties of the original whiting, i.e. they are more readily wetted by organic media.

Precipitated calcium carbonate ($CaCO_3$)

This material is obtained in large quantities as a by-product in water-softening plants when sodium carbonate has been used. It is also produced from the purer grades of limestone (which is essentially $CaCO_3$) by calcining to calcium oxide followed by slaking with water to give calcium hydroxide. Coarse particles are separated and the calcium hydroxide in suspension converted to carbonate by passing carbon dioxide or addition of sodium carbonate solution. The product is filtered, washed and dried.

Properties

Like whiting, it is hydrophilic and is similar to whiting in reactivity; it yields an alkaline aqueous extract and dissolves in dilute hydrochloric, nitric and some organic acids. The physical characteristics are as follows—

Specific gravity 2·7   Oil absorption 24–26
Refractive index 1·6   Mean particle size range 2–4 μm

Uses

The main use is for gloss control in low sheen paints in which it has a stabilizing effect on pigment dispersions. In some paint systems it can be used to control floating and flooding.

Identification tests. As for whiting but no residue is obtained when dissolved in dilute acid.

Coated calcium carbonate

Precipitated calcium carbonate will combine with stearic acid to give a product which is hydrophobic in character. It usually contains about 3 percent stearic acid, the hydrocarbon chains of which are directed away from the particle surface. The result is a particle of lower overall density than the original and of effectively greater diameter. When mixed with pigments which tend to settle to a compact mass the latter is kept softer and more voluminous. It is, therefore, more readily re-dispersed. The amount used should not exceed 10 percent on the total weight of pigment; more than this is likely to impair the flow properties of the paint.

Coated calcium carbonate is marketed under a number of proprietary names [10].

Toxicity of calcium carbonate. The compound is considered to be non-toxic but dusts can be regarded as "nuisance particulates".

TLV (nuisance particulates)   10 mg/m^3.

GYPSUM

This is a crystalline calcium sulphate with the formula $CaSO_4 . 2H_2O$. A purified powdered form was once known as "terra alba" but this term is no longer used.

Gypsum occurs in considerable quantities in Great Britain and is processed by crushing, grinding, washing and drying. The material, however, is used to a very limited extent in paints. The chief interest in gypsum lies in its relationship to Plaster of Paris.

Plaster of Paris

When gypsum is heated to $120° - 130°C$ ($248° - 266°F$), part of the water of crystallization is lost and plaster of Paris is obtained. The latter forms the basis of many commercial plasters used for interior walls. When these plasters are mixed with water, re-hydration sets in with the formation of a dense matrix of gypsum crystals. The reaction is the reverse of that taking place in the formation of plaster of Paris and the two can be represented thus:

$$2CaSO_4 . 2H_2O \underset{\text{Water}}{\overset{\text{Heat}}{\rightleftharpoons}} 2CaSO_4 . H_2O + 3H_2O$$

$$\underset{\text{Gypsum}}{} \qquad \underset{\text{Plaster of Paris}}{}$$

Plaster of Paris is known as a *hemihydrate plaster* and the formula is frequently written as $CaSO_4 . \frac{1}{2}H_2O$.

Calcium silicate

An extender characterized by very fine particle size. It is produced by precipitation from solutions of a calcium salt and an alkali silicate, followed by filtration, washing and drying. The physical characteristics are as follows —

Specific gravity 2·1   Oil absorption 140
Refractive index 1·47   Mean particle size range 30–50 mμ (millimicrons)
pH of aqueous extract 10.

Uses. It is used as a flatting and thickening agent in solvent-based paints. The high pH of the aqueous extract makes calcium silicate an attractive additive to primers for steelwork.

The material is marketed under the trade name "Microcal" [11].

Toxicity. Calcium silicate is classed as non-toxic, but the dust is a "nuisance particulate":

$$\text{TLV} \quad 10 \text{ mg/m}^3$$

Other compounds of calcium used as extenders include

Dolomite, a double carbonate of calcium and magnesium ($CaCO_3MgCO_3$), marketed in the micronized form under the trade name "Microdol" [12].

Calcite, a crystalline form of calcium carbonate, sold under the trade name "Millicarb" [13].

COMPOUNDS OF ALUMINIUM

CHINA CLAY

This is also known as kaolin and is a hydrated aluminium silicate to which the formula $Al_2O_3 . 2SiO_2 . 2H_2O$ is usually assigned.

It is produced by the decomposition of felspar, $K_2O . Al_2O_3 . 6SiO_2$, which is one of the components of granite. The latter consists of a matrix of crystals of quartz, felspar and mica. Breakdown of the granite and subsequent decomposition of the felspar leaves china clay mixed with fairly coarse silica and some mica.

Large deposits of china clay occur in the St Austell area of Cornwall.

Processing

The china clay is quarried by jets of high-pressure water and forms a slurry from which the large pieces of granite and quartz settle out. The slurry is then subjected to levigation to separate silica and mica, and the various grades obtained are filtered, dried and powdered. The grades of finest particle size are often called colloidal china clays. Specially coated grades are also available.

Properties

China clay possesses the following physical characteristics:

Specific gravity	2·6
Oil absorption	30–50 according to grade
Refractive index	1·56
Mean particle size range	1 μm downwards in finer grades.
	Coarser grades 1 μm up to 10 μm

The ultimate particles are flat plates (Fig. 4.2).

It is marketed as a fine white crystalline powder which is practically transparent in organic media. Like many other clay minerals it possesses base-exchange properties. Acid sites are present on the clay particles and neutralization of these sites produces the so-called "alkaline clays". These are characterized by ready dispersibility in water to give deflocculated dispersions.

Both faces and edges of the lamellar clay crystals contain electrically active sites leading to attraction between adjacent particles and the formulation of a flocculated structure. This structure is formed when china clay is used in paint and constitutes one of its most valuable properties.

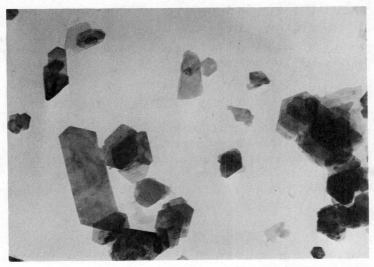

Fig. 4.2   Photomicrograph of China clay (× 18 000)

Uses

The fairly high oil absorption combined with fine particle size makes it a useful flatting and thickening agent for low-sheen paints such as undercoats, eggshell and matt finishes. It is also used in the formulation of some types of "high build" paints. The ability to form a flocculated structure, as described above, is utilized in many paint formulations where settlement is likely to be troublesome. Any settlement is likely to be bulky, soft, and readily reincorporated.

Calcined china clay

Calcination of china clay at 950°–980°C (1742°–1796°F) results in loss of combined water and a fundamental change in crystal structure [14]. The calcined product is harder but less dense than the original material and shows little tendency to induce structure formation in paint systems. During calcination, air voids are created within the particles and these voids confer a degree of light scattering and consequent opacity on the product.

Calcined china clays possess high oil absorption values and are of especial interest for undercoats, flat alkyd enamels, and emulsion paints.

Identification tests

China clay has a soapy feel when rubbed between the fingers. When heated with concentrated hydrochloric acid and evaporated to dryness (preferably several times) it is decomposed. The residue is taken up with dilute hydrochloric acid and filtered. A gelatinous mass of silica remains on the filter and the filtrate gives the normal reactions of aluminium.

Toxicity. The material itself presents no known toxic hazard but the dust can be classed as a "nuisance particulate".

SLATE POWDER

This is an impure hydrated aluminium silicate obtained as a dark grey or greenish grey powder from slate cutting and trimming operations.

It is very hard and coarse and is useful in dark coloured fillers and stoppers because it facilitates rubbing down and does not clog the paper.

Slate powder is cheap and is used as an inert filler in a number of industries.

MICA

This term covers a group of hydrated aluminium silicates which occur in large masses and which are readily split into very thin sheets. These are converted to powder either by dry processing or under water, the latter giving the finer product for paint purposes. The particles are very thin plates and are spoiled by overgrinding. They are sometimes subjected to micronizing and so mica is offered to the industry as dry-ground (coarsest), wet-ground, and micronized (finest).

The type supplied to the paint industry is a hydrated potassium aluminium silicate the composition of which can be represented by the formula $K_2O \cdot 2Al_2O_3 \cdot 6SiO_2 \cdot 2H_2O$, but many variants occur.

Properties and uses

Mica possesses the following physical characteristics:

Specific gravity	2·80–2·85
Oil absorption	75
Refractive index	1·58

It is an inert material and its use in paint is based on the property of the particles to leaf or to orientate themselves parallel with a paint surface and to overlap. This property confers improved moisture resistance and anti-penetrating properties and, in some cases, hard settlement of the pigment in the liquid paint is prevented.

Toxicity. Finely ground mica in the form of dust presents a certain hazard, but this is not as serious as with quartz or asbestos.

TLV (dust)　20 mppcf*　6 mg/m³.

---

* Millions of particles per cubic foot.

BENTONITE

Bentonite is a clay obtained from the United States and consists basically of a hydrated aluminium silicate. It is assigned the general formula $Al_2O_3 . 4SiO_2 . 2H_2O$, but the aluminium is often replaced to varying degrees by magnesium and sodium. The characteristic property of bentonites is their ability to absorb several times their own weight of water and then set to a stiff gel. They are therefore used as thickeners in certain types of water paint.

Bentonites possess the property of undergoing ion exchange reactions with certain organic bases to produce materials which gel in organic liquids, but not in water. These are known as Bentones and are discussed later.

Aluminium hydroxide ("Alumina hydrate")

Aluminium hydroxide releases water vapour when heated to about 200°C (392°F) giving the monohydrate. This reaction is useful in flame-retarding paints in which the compound is sometimes used as an extender. It is marketed in the micronized form.

Aluminium hydroxide is used also as a base for certain organic lakes. In this case it is generally used as a freshly prepared suspension.

COMPOUNDS OF MAGNESIUM

MAGNESIUM CARBONATE

This material occurs naturally as the mineral magnesite, $MgCO_3$, but this form of the carbonate is very seldom used in paint. It also occurs in conjunction with calcium carbonate in the mineral dolomite which is used in the micronized form Microdol [12] as a general purpose extender.

PRECIPITATED MAGNESIUM CARBONATE

The magnesium carbonate used in paint is prepared by precipitation and is a hydrated basic carbonate. It can be manufactured from either magnesite or dolomite and has been represented by the formula $MgCO_3 . 3Mg(OH)_2 . 11H_2O$ (amongst others).

Properties and uses

Precipitated magnesium carbonate possesses the following physical characteristics:

Specific gravity	2·9
Oil absorption	120
Refractive index	1·7

It is a very fine powder which has found use as a flatting agent, especially in undercoats and flat finishes. In recent years it has been displaced to a considerable extent by the hydrated silicas.

Identification tests

On heating it is decomposed and leaves a residue of magnesium oxide, MgO. It is readily soluble in dilute mineral acids with evolution of carbon dioxide and formation of a solution giving the normal reactions for magnesium ions.

ASBESTINE

The source of asbestine is a fibrous mineral occurring in the United States and elsewhere. The asbestine is prepared by dry pulverizing followed by air flotation. It is a hydrated magnesium calcium silicate and has been assigned the formula $Mg_3Ca(SiO_3)_4$, but variations occur in the magnesium to calcium ratio.

Properties and uses

Asbestine possesses the following physical characteristics:

Specific gravity	2·9
Oil absorption	27–30
Refractive index	1·62

The most characteristic property is the fibrous nature of the material and this makes it a good anti-settling agent in paints. It also increases the mechanical strength of certain types of paint film. When being dispersed in paint, and particularly in the ball mill, the fibrous nature of the material (often described as woolly) makes a Hegman gauge reading very difficult to obtain. It is very inert chemically and is not water sensitive.

Identification tests

The fibrous structure is visible under the microscope. Repeated boiling with concentrated hydrochloric acid gives a deposit of gelatinous silica and a solution which gives the normal reactions for calcium and magnesium ions. The silica can be confirmed by drying and by the formation of volatile silicon tetrafluoride when warmed with hydrofluoric acid.

Toxicity

The fibrous nature of the material and the fact that certain types of asbestos have been shown to possess carcinogenic properties have resulted in close attention being paid to the conditions under which asbestine is used. The dust is a "nuisance particulate" and certain types of asbestine [15] are claimed to be physiologically inactive. It is, nevertheless, considered prudent for operatives to wear masks whenever asbestine is used, and to maintain a high standard of industrial hygiene [16].

TALC

This is also known as "french chalk" and, on account of the "greasy" feel, the rock has traditionally been known as "soapstone". It is a hydrated magnesium silicate and can be represented by the formula $Mg_3H_2(SiO_3)_4$ but the ratio of magnesium to silica in commercial deposits can vary considerably. This can be due, in part at least, to the presence of magnesium carbonate (magnesite) in many natural deposits, and this results in a "spread" of properties.

Properties and uses

Talc has the following characteristics:

Specific gravity 2·65–2·85   Refractive index 1·59–1·64
Oil absorption 25–35   pH of aqueous suspension 9–9·5.

Ground talcs contain a variety of particle shapes, but lamellar particles predominate. Particle sizes vary from 5 to 10 μm but micronized grades are available and these possess flatting and anti-settling properties.

Talc is a useful general purpose extender for solvent-based paints. It is hydrophobic and consequently disperses without difficulty. In water, however, talc has a great tendency to flocculate.

Identification tests

The "greasy" or "soapy" feel coupled with examination by microscope is usually sufficient, but the micronized grades are difficult to identify by the latter technique. Warming with dilute hydrochloric acid will dissolve magnesium carbonate (with evolution of CO_2) but decomposition of the hydrated magnesium silicate requires boiling with the concentrated acid. The solution will give the normal reactions for magnesium ions; the silica will remain as an insoluble residue of silicic acid.

Toxicity

As a magnesium compound talc is regarded as non-toxic. The dust, however, is classed as a "nuisance particulate" with TLV 20 mppcf   6 mg/m³.

Note: Care should be taken in the use of "fibrous talc" which may contain a proportion of fibrous tremolite, a material with proven carcinogenic effects in the human body [16].

SILICA

The paint industry uses silica obtained from natural sources as well as manufactured hydrated silicas. The former tend to be very hard and coarse and their use today is limited to certain specific types of product.

NATURAL SILICAS

Quartz

This is a crystalline material and is the commonest form of natural silica, SiO_2. It occurs in many regions of the earth and often in a high state of purity. Processing consists of crushing, grinding and air-classifying.

Properties and uses

Quartz is a very hard crystalline material and the powdered material possesses the following characteristics:

Specific gravity	2·8
Oil absorption	Varies according to particle size
Refractive index	1·53

The finer grades are used in oil-type undercoats, but the main use of the material is in fillers and stoppers. In these the extreme hardness assists the rubbing-down operation.

Flint

The ground form of this material is sometimes used in fillers and stoppers. It is not quite as hard as quartz.

Kieselguhr

This is a member of the family known as diatomaceous earths. These consist of the siliceous remains of diatoms and other creatures which were deposited in primordial seas over long periods of time. The deposit is amorphous, readily powdered and highly absorbent. It is therefore used as a flatting and consistency-control agent in paints and as a filter aid in oil processing.

Celites [17]

These comprise a range of diatomaceous earths and are characterized by excellent colour. They are marketed in various grades ranging from 2 to 20 μm mean particle size, and with oil absorption from 75 to 110. They are used for flatting and for viscosity control in paints.

Diatomite

This is an indigenous diatomaceous hydrated silica obtainable in a range of particle sizes with oil absorptions from 60 to 100. The specific gravity is 1·95 and refractive index 1·46. Diatomite is used as a flatting agent and it is claimed to improve intercoat adhesion.

Toxicity of natural silicas

The properties are controlled by the crystalline form. Amorphous grades are classified as "nuisance particulates", but crystalline quartz is more hazardous and can lead to silicosis.

Manufactured silicas

These are produced by two methods and are sold under a number of trade names [18].

(a) By the action of mineral acid on an alkali silicate solution. A hydrated form of silica (silicic acid) is precipitated and is filtered, washed and dried. This form is amorphous.
(b) By pyrolysis of silicic acid to give an anhydrous silica with a spherical particle shape.

Properties and uses

General physical characteristics:

Refractive index 1·45   Particle size range 3–30 mμ (millimicrons)
Specific gravity 2·0–2·2   (Individual grades lie within this range)
Oil absorption 80–320

These silicas are chemically inert and by virtue of their large surface area they are used on a considerable scale as flatting and viscosity-control agents in solvent-based paints. They are very pure and make no contribution to either colour or opacity.

Toxicity of manufactured silicas. As non-crystalline materials these are classified as "nuisance particulates" — see "Toxicity of natural silicas" (*above*).

References to Chapter 3

[1] MURLEY, R. D., Some aspects of the physical chemistry of titanium pigment surfaces. *JOCCA*, 1962, **45**, 16.
[2] KENNEDY, D. R. *et al.*, *Trans. Far. Soc.*, 1958, **54**, 119.
[3] MYERS, R. R., Report of the Research Director, Paint Research Institute. *J. Coatings Tech.*, 1978, **50**, No. 640, 51.
[4] Tioxide International Ltd.
[5] CLAY, H. F. and COX, J. H., Chromate and phosphate pigments in anti-corrosive primers. *JOCCA*, 1973, **56**, 13.
[6] National Lead Industries Inc.
[7] DAVIDSON, S. L., A new versatile lead-free pigment. *JOCCA*, 1975, **58**, 435.
[8] Sachtleben Chemic. G.m.b.H.
[9] Associated Lead Manufacturers Ltd.
[10] "Calofort" (J. & E. Sturge, Ltd); "Winnofil" (ICI Ltd); "Polcarb" (ECC International Ltd).
[11] Joseph Crosfield & Sons Ltd.
[12] Norwegian Talc Industries.
[13] Melbourn Chemicals Ltd.

[14] Publication R 63, "Extenders for the Paint Industry". ECC International Ltd.
[15] "Lubestine" (Bromhead & Dennison Ltd).
[16] Asbestos Regulations, 1969. H.M. Factory Inspectorate Technical Data, Notes 12 and 13.
[17] Johns-Manville, Denver, Colorado.
[18] "Gasil" (Joseph Crosfield & Sons Ltd); "Aerosil" (Degussa); "Cab-o-Sil" (Cabot).

5 Yellow, orange, and red inorganic pigments

The important pigments in this range are the chromes (lead, zinc, barium and strontium); the cadmium colours; the yellow and red oxides of iron; and red lead. Some interesting but lesser-used pigments are also described.

THE CHROME PIGMENTS

Lead chromes (CI Pigment Yellow 34, Pigment Red 104)

These all contain lead chromate, either alone or co-precipitated with other lead salts. The colours range from pale primrose to orange and scarlet. The following table shows the relationship between colour and composition.

Lead chrome	*Composition*
Primrose	Lead chromate, lead sulphate and alumina
Lemon	Lead chromate and lead sulphate (relative proportions determine the shade)
Middle	Lead chromate only
Orange	Lead chromate and lead hydroxide
Scarlet	Lead chromate, lead molybdate, and lead sulphate

Manufacture of lead chromes

They are all produced by precipitation using solutions of lead salts and solutions of chromates or mixtures of chromate with sulphate or molybdate.

The lead salts employed are made from litharge by conversion to a solution of nitrate or acetate, but basic salts such as basic acetate or basic chloride are sometimes used in the form of suspensions. Lead nitrate is generally preferred to the acetate because it gives chromes that are faster to light.

Sodium dichromate is usually employed as the source of chromate. It is cheaper than the potassium salt and in addition is more deliquescent and consequently easier to dissolve. Before addition to the lead solution it is converted to the normal chromate by treatment with sodium carbonate solution:

$$Na_2Cr_2O_7 + Na_2CO_3 \rightarrow 2Na_2CrO_4 + CO_2$$

For the primrose and lemon chromes sulphuric acid or a sulphate such as sodium sulphate is added at the same time as the chromate solution. This results in co-precipitation of the lead chromate and sulphate, giving a brighter and greener pigment than would result from a mixture of the two.

The conditions of precipitation must be rigidly controlled. The solutions employed are usually 10 percent and precipitation is carried out at normal temperature. A slight excess of lead is left in solution at the end to give the brightest pigment, and for the paler shades the pH is generally about 5.

Particle shape

The lead chromes are crystalline pigments but the form of crystal is not constant. It has been established that the crystal form has a profound influence on the properties of the pigment. The paler shades usually precipitate in the rhombic form but change to the monoclinic on standing. In the latter form the pigments are more durable and fast to light. With the primrose chrome, the greenish shade is obtained by stabilizing the rhombic form by the addition of a little alumina.

Primrose chrome is the palest and greenest of the lead chromes. The greenish shade is obtained by using alumina to stabilize the rhombic form in which it is precipitated. Tartaric acid also is used as a stabilizer.

Lemon chrome. The depth of colour of this pigment can be adjusted by varying the ratio of lead sulphate to lead chromate in the precipitate. An increase in lead sulphate results in a paler product. Lemon chromes are usually supplied in pale, normal and deep shades. These shades are redder than the primrose and possess a monoclinic crystal form.

Mid chrome. This is the normal lead chromate, $PbCrO_4$, which has a monoclinic crystal form. It is deeper and redder than the lemon chromes.

Orange chrome is a basic lead chromate the shade of which can be varied by altering the ratio of lead chromate to lead hydroxide.

Scarlet chrome. If lead chromate is co-precipitated with lead molybdate the products are scarlet or deep orange pigments of exceptional brilliance and opacity. Often a proportion of lead sulphate is precipitated at the same time and improves the colour and stability of the pigment. The composition of a scarlet chrome is approximately 80 percent lead chromate, 12 percent lead molybdate and 8 percent lead sulphate. Scarlet chrome is supplied in three depths of shade, resulting from variations in the ratios of the ingredients.

Properties of lead chromes

There is an increase in depth and redness of colour in passing from the greenish-yellow primrose through lemon and mid to the very reddish orange chrome. Scarlet chrome occupies a rather unique position. It is brighter and cleaner than the orange chromes and possesses much greater tinting strength.

The lead chromes possess the following physical characteristics:

Specific gravity	5·9–6·2
Oil absorption	12–16
Refractive index	2·4

They are all bright pigments with good opacity and high tinting strength.

Their general durability is good, but in the past they have shown a tendency to darken slightly on long exterior exposure in paint films. Present-day lead chromes are virtually free from this defect. When exposed in urban atmospheres, however, they are darkened by the presence of hydrogen sulphide and tend to bleach when exposed to sulphur dioxide.

Lead chromes are non-reactive toward paint media and compatible with most other pigments. However, it is considered inadvisable to mix them with pigments containing sulphides such as ultramarine in view of the risk of darkening through lead sulphide formation. When mixed with Prussian Blue they form chrome greens (Chapter 6).

Toxicity of lead chromes

As compounds of lead, these pigments are classed as toxic and their use is controlled by the Lead Paint Regulations. Additionally they present the chromate hazard, and the TLV for airborne dust, calculated as CrO_3, is 0.1 mg/m^3.

On account of their toxic properties, lead chromes and other lead pigments are no longer used in paints for children's toys and for the same reason, they have been phased out of decorative paints.

They continue to be used in some types of industrial maintenance paints where their toxicity is not likely to constitute a hazard (except, possibly, when they are eventually removed). If the paints are to be applied by spray, the Regulations lay down a maximum of 5 percent "soluble lead" calculated as PbO on the pigment. This is lead which is soluble in 0.25 percent hydrochloric acid (the concentration of this acid which is present in the stomach). The method of determination is described in BS 282:1963 [1], but manufacturers of lead chrome usually declare the figures for their products.

It is probable that in the foreseeable future all paints will be based on leadless pigments.

Basic lead silicochromate

An interesting pigment consisting of a complex lead chromate/lead silicate which is produced when lead chromate is calcined on to a silica core. It is marketed in two grades, differing in particle size, under a trade name [2].

Properties and uses

Specific gravity $4.0–4.1$ Oil absorption $13–18$.

The pigment is of a dull orange colour and can be used in a wide range of media. As a lead compound it is toxic and its use is controlled by legislation (see Toxicity, above). The principal use of the pigment is in corrosion-inhibiting primers for steelwork. It is claimed to be effective on steel from which rust has not been totally removed.

Identification test

On heating with dilute hydrochloric acid it is partly soluble, giving a solution containing lead and chromate ions (on long boiling the chromate may be reduced to a chromic salt with evolution of oxides of chlorine—these are highly toxic and must not be inhaled).

Zinc chromes (CI Pigment Yellow 36 & 36.1)

The zinc chromes in common use comply with the requirements of BS 282:1963 [1] in which three types are described:

Type 1　A zinc potassium chromate with no specified limits on impurities. It is therefore not suitable for use in metal primers, which is the main use of zinc chromes.

Type 2　A zinc potassium chromate, approximating to the formula $3ZnCrO_4 . K_2CrO_4 . Zn(OH)_2$. It contains an appreciable quantity of soluble chromate but is free from other salts.

Type 3　Zinc tetroxychromate, $ZnCrO_4 . 4Zn(OH)_2$. Contains very little soluble chromate and is free from other salts.

Type 2: Zinc chrome

A greenish-yellow pigment with poor opacity.

Specific gravity 3·4–3·5　　　　Oil absorption 14–18.

It dissolves readily in dilute mineral acids and reacts with many organic acids.

The main use of zinc chrome is in primers for steel and for light alloys. The protection of steel is thought to result from the presence of soluble chromate in the primer film. Moisture permeating the film carries chromate ions to the steel surface when they polarize the anodic areas of potential corrosion cells. Primers used for this purpose usually contain iron oxide in addition to zinc chrome to give the primer the necessary opacity.

A different mechanism, although also dependent on the presence of soluble chromate, is thought to operate in primers applied to aluminium and aluminium alloys. Unpainted aluminium carries a film of adherent oxide which protects the metal (except under severe conditions). The oxidizing action of the chromate maintains the oxide film and repairs any damage.

Identification tests. Zinc chrome is soluble in dilute acids and alkalis. The solution in acid gives the normal reactions for zinc, potassium and chromate ions. When the pigment is warmed with water a yellow "bleed" (soluble chromate) is apparent.

Type 3: Zinc tetroxychromate

This resembles the Type 2 material in colour and poor opacity. It also is soluble in dilute mineral acids and reacts with a number of organic acids. The soluble chromate content, however, is very low and is irrelevant to the principal use of the pigment in etch primers. Here it is dispersed in an alcoholic solution of polyvinyl butyral resin. Phosphoric acid is added to confer the "etching" properties, but film formation from etch primers is a complex process.

Identification tests. Zinc tetroxychromate shows less "bleed" in water than does the Type 2 pigment. Like the latter it is soluble in dilute acids and alkalis. The solution in acid gives the normal reactions for zinc and chromate ions.

Toxicity of zinc chromes

The hazards are associated with the chromate radical, and great care must be taken in handling soluble dichromates and chromic oxide, CrO_3, during the manufacture of zinc chrome. Skin ulcers as well as attack on the nasal membranes can result from contact. A hazard also arises from zinc chrome dust, and adequate protection must be provided when handling the pigment.

$$\text{TLV of chromates (as } CrO_3) = 0\cdot1 \text{ mg/m}^3$$

The Toys (Safety) Regulations, 1974 [3] lay down a limit of 250 parts "soluble" chromium per million parts of dry paint film.

Barium chromate (CI Pigment Yellow 31)

This pigment has the formula $BaCrO_4$ and can be prepared by precipitation from solutions of barium chloride and a chromate. It is a very weak, greenish-yellow pigment and is used in stoppers for use on light alloys.

It is soluble in dilute hydrochloric acid, the solution giving the usual reactions for barium and chromate ions.

Toxicity

Soluble barium compounds are toxic and the ready solubility of barium chromate in dilute hydrochloric acid places the pigment in the "toxic" class. $TLV = 0\cdot5$ mg/m³, but exposure to this concentration of dust should not exceed 15 minutes.

Strontium chromate (CI Pigment Yellow 32)

Strontium chromate is a brighter colour than the barium compound. It has been found very suitable for use in water-soluble stoving primers for application by spray, dip or electrodeposition.

Identification tests. Strontium chromate dissolves in dilute hydrochloric acid on warming to give a solution containing strontium and chromate ions.

Toxicity. The hazards associated with the manufacture and use of strontium chromate are associated with the chromate radical and are similar to those associated with zinc chrome (*q.v.*).

CADMIUM COLOURS (CI Pigment Yellow 37, Pigment Red 108)

The term covers a number of colours ranging from a pale primrose to lemon, orange, scarlet, crimson and maroon. The paler colours consist of cadmium sulphide, but the deeper shades are co-precipitated cadmium sulphide and cadmium selenide.

Cadmium sulphide exists in two crystalline forms, the α form which is yellow and the β form which is red. Intermediate shades between the two extremes, and which consist of mixtures of the two forms, can be obtained by varying the conditions of precipitation, i.e. concentration, temperature, types of salts used, and pH.

Manufacture

The basic reaction is treatment of a cadmium salt solution with a sulphide such as hydrogen sulphide:

$$CdSO_4 + H_2S \rightarrow CdS + H_2SO_4$$

or sodium sulphide, which must be freshly prepared:

$$CdCl_2 + Na_2S \rightarrow CdS + 2NaCl$$

It is essential that no free sulphur is present during the precipitation as this reduces the stability of the pigment.

Washing must be very thorough since traces of soluble salts exert an adverse effect on performance.

The pigments are dried at about 80° to 85°C (176° to 185°F) and then calcined in an inert gas at about 500°C (932°F) to develop the colour and tinting strength.

Primrose shade

This is the palest shade and contains about 5 percent zinc sulphide co-precipitated with the cadmium sulphide.

Pale lemon shade

This is the α form of cadmium sulphide. A very dilute solution of cadmium chloride is used and precipitation carried out near 0°C (32°F). A slight excess of cadmium at the end-point gives the brightest product and improves stability.

Deep lemon shade

This is the β (redder) form of cadmium sulphide. This also is made from cadmium chloride using more concentrated solutions and higher temperatures. A greater excess of cadmium should be present at the end-point.

Orange-maroon shades

These consist of co-precipitated cadmium sulphide and selenide and are obtained by treating a cadmium salt solution with a freshly prepared solution of a mixture of sodium sulphide and sodium selenide:

$$2CdCl_2 + Na_2S + Na_2Se \rightarrow CdS.CdSe + 4NaCl$$

The different shades are obtained by changing the ratio of sulphide to selenide; generally in the orange and scarlet shades the ratio of CdS to CdSe is approximately 100 to 12, and in the maroons it is about 100 to 40.

Properties and uses

The cadmium colours possess the following physical characterstics:

	Yellows	Deep shades
Oil absorption	32–35	22–28
Specific gravity	4·2–4·3	4·5–4·7

They are bright colours, although the red shades are a little dull compared with the majority of red organic colours. Although attacked by concentrated mineral acids they are resistant to organic acids, dilute mineral acids and alkalis. They are non-bleeding in paint media, possess good opacity, and are stable to heat up to 500°C (932°F). The fastness to light is good, but fading on weathering has been observed with tinted whites in alkyd/amino stoved finishes.

Cadmium pigments are expensive and this limits their use as replacements for the toxic lead chromes. Their main use is in heat-resisting paints.

Identification tests

Cadmium colours are resistant to heat and alkali but are decomposed on heating with concentrated hydrochloric acid. Hydrogen sulphide or a mixture of hydrogen sulphide and hydrogen selenide is evolved and recognized by the characteristic smell. The solution gives the normal reactions for cadmium ions.

Cadmium lithopones (Cadmopones)

These materials bear some resemblance to the normal lithopones and are prepared in a similar way by double decomposition using solutions of cadmium sulphate and barium sulphide.

They behave as reduced forms of the parent cadmium colours and their only virtue lies in lower cost which, of course, is accompanied by lower tinting strength and opacity.

Toxicity

The solubility of cadmium pigments is very low but, in handling these materials, care must be taken to avoid inhalation of dust. Masks should be worn.

A danger arises in welding through films or burning-off paint containing cadmium pigments. This can lead to the formation of toxic cadmium oxide, CdO.

TITANIUM NICKEL YELLOW

This is a very interesting member of the rather limited range of heat-resisting pigments.

It is prepared from the titanium hydroxide precipitate produced in the sulphate process (Chapter 5). The slurry is placed in a vat, treated with salts of nickel and antimony, filtered and calcined. The product is ground and graded.

Properties and uses

The pigment is a compound of titanium, nickel and antimony with specific gravity 4·5 and oil absorption 22 to 25. It has a dullish yellow self-colour but gives very bright and clean tints on reduction with white. The fastness to light is very good, even in the pale tints. It is also fast to solvents, paint media, acids and alkalis.

A very valuable property is its fastness to heat; it is stable up to 500°C (932°F). In this respect it rivals the cadmium yellows.

The principal use of this pigment is as a stainer for heat-resisting paints.

YELLOW AND RED OXIDES OF IRON

NATURAL OXIDES

Iron oxide is one of the most abundant mineral oxides in the earth's crust. It is widely distributed and has been used as a pigment for many centuries. If the deposits are to be economically workable as pigments they must be of good colour, uniform, free from contamination, and must contain a reasonable quantity of iron oxide. The colours of the natural oxides range from yellow to black, and the impurities are usually silica together with compounds of aluminium and calcium.

Principal ores of iron

(i) *Limonites* or hydrated oxides in which the constitution can vary from $2Fe_2O_3 . H_2O$ to $Fe_2O_3 . 4H_2O$. These comprise the yellow ochres, the siennas and the umbers.

(ii) *Haematite*, which consists basically of Fe_2O_3 and constitutes the red oxides. It is the principal source.

(iii) *Magnetite* (or *lodestone*). A magnetic black oxide with formula Fe_3O_4.

(iv) *Micaceous oxide of iron*. So called because the particles are lamellar, i.e. in the form of thin plates, and in this respect the particles resemble those of mica. It is a greyish-black colour.

(v) *Siderite* or *Spathic* ore. Mainly ferrous carbonate and used for smelting.

(vi) *Iron pyrites*, FeS_2, used as a source of sulphur dioxide in sulphuric acid manufacture.

The hydrated oxides
Ochres

Ochres are actually coloured clays and consist of mixtures of hydrated oxide of iron with mineral silicates. The composition and colour vary considerably from one source to another. Generally the colour is described as yellow but is dirtier and redder than the yellow of the paler lead chromes or of the cadmium sulphides.

Ochres are widely distributed, and good quality materials are produced in France and South Africa, among other places.

Processing

This follows the normal pattern for natural pigments — grinding, washing, drying, disintegrating and grading. Much of the yellow ochre supplied to the paint industry today has been micronized.

Properties and uses

In iron oxide pigments, properties such as tinting strength and opacity vary with the iron oxide content. The latter figure for ochres can range from 16 to 60 percent and there is therefore a very wide variation in pigment properties. The specific gravities average 2·8 to 3·0 but the oil absorptions range from 20 to 40.

Much of the ochre used by the paint industry today is the micronized variety and is used as a stainer for gloss paints, undercoats and emulsion paints when it confers a pleasant reddish cream tint. It is, however, being displaced by the stronger but yellower Ferrite Yellow.

Siennas

These are brownish yellow in colour and contain a small quantity (about 1 percent) of manganese dioxide. They are obtained principally from Italy and Sicily. Like the ochres, siennas vary in composition, the iron oxide content ranging from 40 to 70 percent.

Siennas are usually micronized and in this form are used as stainers. They are semi-transparent in oil and were much used in scumble stains when graining was more fashionable.

The material in the condition as mined is known as Raw Sienna. If this is calcined, dehydration of the hydrated iron oxide takes place to produce the orange-red Burnt Sienna. The physical characteristics of the two forms are as follows:

	Raw Sienna	*Burnt Sienna*
Specific gravity	3·1–5·0	3·5–5·0
Oil absorption	35–45	35–45

Umbers

These are dark brown in colour and contain up to 15 percent of manganese dioxide. The best grades have been obtained from Cyprus, but the composition varies, the iron oxide content ranging from 30 to 50 percent.

They are semi-transparent in oil and were formerly valued as pigments for scumble stains. The micronized forms are sometimes used as paint stainers.

Like the siennas, the umbers are marketed in two grades, Raw and Burnt, the latter being a very rich reddish-brown. Their specific gravities range from 3·5 to 4·5 and oil absorption from 32 to 50.

Red oxides

Deposits of ferric oxide of varying quality occur in many parts of the world, but few English deposits are suitable as pigments.

Extensive deposits occur in Spain from which considerable quantities are imported into this country. The ground, levigated and dried ore contains 80 to 90 percent of iron oxide and is frequently blended with other grades for the market. Spanish oxide is a bright colour and has been popular in paint manufacture for many years.

A bluer shade of natural red oxide is provided by the Persian Gulf material known, in short, as Gulf Red. This is a very bright rich colour and was formerly known as Indian Red, but this term is now reserved for certain grades of manufactured red oxides.

A fault of the natural oxides is lack of uniformity, and so pigment manufacturers offer blends which are consistent in quality and, by micronizing, the texture and particle size are also closely controlled.

The natural oxides vary in specific gravity from 3·5 to 5·0 and the oil absorption ranges from 12 to 30.

Identification tests

The natural yellow and red oxides of iron dissolve on boiling in concentrated hydrochloric acid to give solutions with the normal reactions for ferric ions. With the siennas and umbers the solutions also contain some manganese ions. In all cases there is an insoluble siliceous residue, which is absent from manufactured oxides.

SYNTHETIC YELLOW AND RED OXIDES

There are two general processes used for the manufacture of these pigments — precipitation and calcination. In each case the basic raw material is crystalline ferrous sulphate, $FeSO_4 . 7H_2O$, known as copperas or green vitriol. This is obtained as a by-product in the tin-plate and galvanizing industries where the steel sheets are pickled in sulphuric acid.

Another source of ferrous sulphate is the sulphate process for the manufacture of titanium dioxide from ilmenite (Chapter 4). Large quantities are obtained as a by-product from this process.

Precipitation process

The following oxides are produced by this process.

Ferrite yellow (CI Pigment Yellow 42)

Also known as yellow oxide of iron. It is manufactured in the following way. A solution of ferrous sulphate is treated with a solution of an alkali, e.g. sodium hydroxide, to precipitate ferrous hydroxide:

$$FeSO_4 + 2NaOH \rightarrow Fe(OH)_2 + Na_2SO_4$$

The ferrous hydroxide is then oxidized by bubbling air through the suspension:

$$4Fe(OH)_2 + O_2(air) \rightarrow 2(Fe_2O_3 . H_2O) + H_2O$$

The formula $Fe_2O_3 . H_2O$ is often assigned to ferrite yellow, but the form $FeO.OH$ is more commonly used. However, the chemical composition and colour of the pigment can vary with the concentration of the solutions, the temperature of precipitation, and the temperature and length of time for which air is passed. After oxidation the liquor containing the precipitate is boiled to coagulate the pigment and to improve the colour. The pigment is filtered, washed very thoroughly to remove soluble salts, and dried at $60°$ to $65°C$ ($140°$ to $149°F$. It is then powdered.

Properties and uses

Ferrite yellow (Fig. 5.1) is a soft pigment with specific gravity 4·0 and oil absorption 50 to 60. It is cleaner and brighter than ochre and possesses much greater tinting strength as a result of the higher iron oxide content (about 90 percent as Fe_2O_3). It has excellent lightfastness and is completely insoluble in oils, resins and solvents. On heating it reddens as a result of loss of combined water and is converted to red oxide, Fe_2O_3.

It is widely used as a stainer in gloss paints, undercoats and aqueous paints.

Identification

On strong heating it is converted to red oxide, Fe_2O_3. It is soluble in hydrochloric acid to give a clear solution and leaves no residue. The solution gives the usual reactions for ferric ions.

Fig. 5.1   Photomicrograph of yellow iron oxide (× 15 000)

Red oxide (CI Pigment Red 101)

This can be manufactured in the following way. A solution of ferrous sulphate is treated with a solution of sodium carbonate to precipitate ferrous carbonate:

$$FeSO_4 + Na_2CO_3 \rightarrow FeCO_3 + Na_2SO_4$$

The suspension is heated and air is blown through to oxidize the ferrous carbonate to ferric oxide:

$$4FeCO_3 + O_2 \rightarrow 2Fe_2O_3 + 4CO_2$$

The addition of 2 to 3 percent of magnesium carbonate gives a brighter, cleaner and stronger grade of red oxide.

The shade of the product is influenced by the temperature and duration of the air blowing. Thus at 60°C (140°F) an air blow of 24 hours gives a yellow shade oxide; at 95°C (203°F) an air blow of 18 to 20 hours produces a blue shade pigment.

The pigment is filtered and washed very thoroughly to remove all traces of soluble salts. This is important as many red oxides are used in primers for steelwork where soluble sulphates (or chlorides) act as corrosion catalysts. The oxide is then dried and ground (Fig. 5.2).

Properties and uses

Compared with the natural red oxides, the synthetic type is brighter,

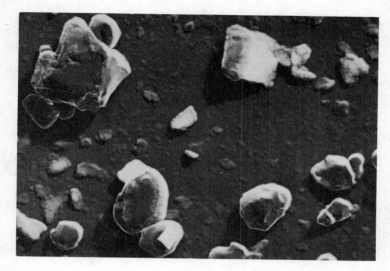

Courtesy: Paint Research Association

Fig. 5.2   Photomicrograph of red iron oxide (× 30 000)

cleaner and has a higher tinting strength due to higher Fe_2O_3 content (over 99 percent). The opacity, which is also proportional to the ferric oxide content, is also higher.

The pigment has a specific gravity of 5·0 and an oil absorption of 30. It is fast to light and insoluble in oils, solvents, and resins. It is resistant to all alkalis and dilute acids but attacked by hot concentrated acids. It is also heat resistant.

Synthetic iron oxide is used in air-drying and stoving finishes, undercoats and primers, emulsion paints, plastics, and cements.

Identification tests

Boiling with concentrated hydrochloric acid gives a clear solution showing the usual reactions of ferric ions. It leaves no residue and so differs from the natural material.

Calcination process

This process also utilizes ferrous sulphate crystals, $FeSO_4 \cdot 7H_2O$ as the basic raw material and is used, in the main, for the deeper shades of iron oxide. The ferrous sulphate crystals are dried in a heating chamber and are thereby converted to the monohydrate, $FeSO_4 \cdot H_2O$. The monohydrate is then fed into a rotary calciner with air and sulphur dioxide. The latter prevents the formation of basic salt and regulates the reaction. The temperature employed is 600°C rising to 800°C (1112° to 1472°F) and the sequence of reactions can be represented by the equations:

$$FeSO_4 . 7H_2O \xrightarrow{\text{Heat}} FeSO_4 . H_2O + 6H_2O$$

$$6FeSO_4 . H_2O \xrightarrow{\text{Calcine}} 2Fe_2O_3 + Fe_2(SO_4)_3 + 6H_2O + 3SO_2*$$

$$Fe_2(SO_4)_3 \xrightarrow{\text{Calcine}} Fe_2O_3 + 3SO_3$$

$$* 2SO_2 + O_2(\text{air}) \rightarrow 2SO_3$$

The frit from the calciner is discharged into water when it breaks up. It is washed thoroughly with hot water to remove traces of acidity, dried and ground.

This process produces the yellowest and strongest of the calcination oxides known as *Turkey Red*.

If the temperature of calcination is increased to 900°C (1652°F), the pigment produced is deeper, bluer and weaker than Turkey Red. It is known as *Indian Red*.

At still higher temperatures (up to 1000°C, 1832°F) the product is still deeper and is known as *Purple Oxide* or *Purple Brown*.

Properties and uses

The calcination oxides possess specific gravities of 4·5 to 5·0, and the oil absorptions range from 20 to 25.

Turkey Red is a bright soft pigment with good opacity and tinting strength. Indian Red is deeper, bluer and harder and possesses lower opacity and tinting strength. These properties are developed still further in the Purple Oxides.

These pigments are sometimes used in gloss paints and undercoats. In primers, however, their use is somewhat risky due to the possibility of the presence of sulphate which is extremely difficult to remove by washing. They are fast to light and heat and are insoluble in all oils, resins and solvents. They are also resistant to alkali and acids except boiling concentrated acids, and even here the attack is slow.

Identification tests

Boiling (usually prolonged) with concentrated hydrochloric acid yields a solution showing the usual reactions for ferric ions.

Bauxite residue

The mineral Bauxite, which is one of the chief sources of aluminium, contains appreciable quantities of iron oxide. After extraction of the aluminium, the residue is ground and sold under the name "Bauxite Residue". It contains 60 to 70 percent Fe_2O_3, the balance consisting of siliceous matter and possibly traces of alumina.

It is sometimes used in primers for steelwork, as the protective properties appear to be superior to some of the natural red oxides.

RED LEAD (CI Pigment Red 105)

Manufacture

Red lead is manufactured by calcining litharge (PbO) in an oxidizing atmosphere at about 350°C (662°F). Reaction is slow but cannot be accelerated since decomposition of the red lead takes place below 500°C (932°F). Accurate temperature control is therefore necessary. The reaction can be represented by the equation:

$$6PbO + O_2 \underset{470°C}{\overset{350°C}{\rightleftharpoons}} 2Pb_3O_4$$

The final product always contains some free litharge, the amount of which can be controlled by the degree of oxidation.

Properties

Red lead is an orange-red pigment with the following physical characteristics:

Specific gravity	8·8
Oil absorption	7–8
Refractive index	2·42
Mean particle size range	3–12 μm

The pigment is marketed in three grades conforming to the requirements of BS 217 [4]. These are:

Type 1   Non-Setting Red Lead
Type 2   Ordinary Red Lead
Type 3   Red Lead for Jointing Purposes

The three types differ in respect of the free PbO content, and the following figures show the composition of some of the present-day red leads:

	Type 1	*Type 2*	*Type 3*
Total oxides of lead	Not less than 99·5%	not less than 99·5%	not less than 99·5%
Pb_3O_4	Not less than 95%	not less than 72% and not more than 90%	not less than 43% and not more than 72%
Free PbO	Balance	Balance	Balance

The free PbO content determines the use of the pigment. A high proportion, as in Type 3, will react rapidly with linseed oil and set to a hard mass. Type 2 reacts more slowly, but paints made with this type can be stored for no longer than 24 hours. Type 1 is used in paint manufacture and gives stable paints with good working properties.

Red lead is a reactive pigment and its chemical reactions support the accepted constitution $2PbO.PbO_2$. When dispersed in linseed oil or other oxidizing media lead soaps are formed by reaction between the PbO and acidic by-products produced during the oxidation process. These soaps are thought to play an important part in the protection afforded to steelwork by red lead primers. The traditional medium for red lead primers has been linseed oil, but although the protective qualities have been excellent, the primers were slow-drying and remained soft for a considerable time. Other quicker-drying media have been used, including alkyds, epoxy esters, and chlorinated rubber.

Red lead is a very toxic pigment, its manufacture and use being controlled by the Lead Paints Regulations, 1927, and subsequent amendments. As a result of this high toxicity the use of the pigment is steadily declining, a process which has been accelerated by the introduction of the non-toxic zinc phosphate.

Identification tests

When strongly heated, red lead decomposes and then melts to an orange liquid which sets to a yellow solid (litharge). In the decomposition oxygen is evolved and the reaction is the reverse of the process of manufacture.

On treatment with dilute nitric acid the PbO portion of the molecule is dissolved out, leaving a brown residue of PbO_2. Addition of a few crystals of a reducing agent such as sodium nitrite or oxalic acid reduces the PbO_2 and gives a clear solution of lead nitrate. This gives the normal reactions for lead ions.

LEAD CYANAMIDE, $PbCN_2$

This is a relatively modern pigment which was produced by Hoechst in Germany at the end of the 'twenties.

Manufacture

Calcium cyanamide is stirred with sodium carbonate solution:

$$CaCN_2 + Na_2CO_3 \rightarrow CaCO_3 + Na_2CN_2$$

The solution of sodium cyanamide is then stirred with lead sulphate:

$$Na_2CN_2 + PbSO_4 \rightarrow PbCN_2 + Na_2SO_4$$

The pigment is filtered, washed and dried.

Properties and uses

Lead cyanamide is a yellow pigment with specific gravity 6·8 and oil absorption 21 to 23. The extract obtained on treating 5 g of the pigment with 100 ml of distilled water is alkaline, the pH being 9·3.

It is a very reactive pigment and forms lead soaps with drying oils. When made into paints it does not settle in the tin and the paints show good brush-ability.

The pigment is very effective in corrosion-inhibiting primers for steelwork and this is probably due, in part at least, to its highly alkaline nature. It has also been used in wood primers and undercoats and can be used in drying oil, oleoresinous or alkyd media.

As a lead compound it is, of course, subject to the Lead Paint Regulations.

References to Chapter 5

[1]   BS 282:1963. Lead chromes and zinc chromes for paints. British Standards Institution, 2 Park Street, London.
[2]   "Oncor". National Lead Industries Inc.
[3]   H.M. Stationery Office. See also BS 3443:1968, Code of safety requirements for children's toys and playthings. British Standards Institution.
[4]   BS 217:1961. Red lead for paints and jointing compounds. British Standards Institution.

6 Blue and green inorganic pigments

BLUE PIGMENTS

There are only three pigments of importance in this class. These are Ultramarine, Prussian and Cobalt blues.

Ultramarine Blue (CI Pigment Blue 29)

The natural material, a mineral known as "lapis lazuli", occurs in Iraq, Tibet and China and has been used as a pigment for many centuries. It was prized for its brilliance and purity of tone and although the use of lapis lazuli as a paint pigment was discontinued many years ago, it has continued to be used in artists' colours until very recent years.

Lapis lazuli possesses a very complex constitution containing silicates of lime, alumina and sodium together with sulphides of the latter.

Early in the nineteenth century several observers reported a blue compound in certain blast furnace slags and this was found to give similar reactions to lapis lazuli. This observation stimulated further research which led ultimately to the modern process for the manufacture of Ultramarine.

Manufacture

The raw materials employed include sulphur, soda ash, siliceous materials such as china clay, silica, etc., and a carbonaceous material such as rosin. These are mixed and heated in a muffle furnace for some days. The product is ground, washed, dried and graded.

Variations in shade of blue are obtained by altering the proportions of the ingredients. Rose and violet shades are produced by treatment of the blue pigment with hydrochloric acid gas at about 275°C (524°F) [1].

Properties and uses

Ultramarine Blue is a bright pigment and is one of the reddest blues available; the colour is not easily matched with other pigments. It has the following physical characteristics—

Specific gravity 2·35   Refractive index 1·51
Oil absorption 30–35   Mean particle size diameter 0·1–3·0 μm

As a consequence of the method of manufacture, the heat resistance is excellent. It is insoluble in all paint media and solvents, is fast to alkali (except in concrete), but is readily attacked by mineral acids.

Ultramarine Blue possesses good lightfastness, but loss of colour can occur if it is exposed in a paint film to a moist acidic atmosphere. It is strongly hydrophilic and disperses readily in water, but it can migrate into water from oil or other organic medium. "Ultra blue" is therefore used mainly in aqueous media, but it is also employed in organic media where a cationic or non-ionic surfactant assists wetting and helps to stabilize the dispersion. It is widely used in screen inks.

The constitution of Ultra blue is related to that of the zeolites and ion-exchange reactions are possible. By the action of silver nitrate, for example, sodium in the blue can be replaced by silver to give a yellow-silver ultra-marine.

Toxicity. Ultramarine Blue possesses no known toxic hazard, but the dust can be classed as a "nuisance particulate".

Identification tests

The pigment is unchanged on ignition or on treatment with alkalis. It is decomposed by dilute mineral acids, giving hydrogen sulphide and a pre-cipitate of sulphur and leaving a siliceous residue. This test differentiates between Ultramarine and Prussian Blue, the latter being decomposed by alkali but not by dilute acids.

PRUSSIAN BLUES (CI Pigment Blue 27)

These are iron compounds of the ferrocyanide radical $[Fe(CN)_6]''''$ and contain also an alkali metal or ammonia. They are known in the United States as Iron Blues. In this country there are two types known as Potash or Non-bronze Blues and Ammonia-soda or Bronze Blues.

Potash or non-bronze blues

These are usually considered to be forms of ferric potassium ferro-cyanide, $KFe[Fe(CN)_6]$ and can be prepared by either the Indirect or Direct process. These blues precipitate in a very finely divided condition which renders them very difficult to filter. They are therefore made in as coarse a condition as possible by precipitation at or near $100°C$ ($212°F$).

Indirect process of manufacture

A hot 10 percent solution of potassium ferrocyanide is added slowly, with stirring, to a hot ($95°$ to $100°C$, $203°$ to $212°F$) 10 percent solution of ferrous sulphate containing 1 percent sulphuric acid. A white precipitate of ferrous potassium ferrocyanide is formed according to the equation:

$$FeSO_4 + K_4Fe(CN)_6 \rightarrow FeK_2 . Fe(CN)_6 + K_2SO_4$$

The precipitate rapidly turns blue on the surface as a result of oxidation by the oxygen of the atmosphere.

The oxidation to the blue ferric potassium ferrocyanide can be carried out by a number of oxidizing agents such as chlorates, dichromates, and chlorine. Air can be used, provided the suspension is kept sufficiently acid to guard against formation of ferric oxide. Potassium chlorate is widely used, the solution being kept acid with sulphuric acid. Under these conditions the oxidation can be represented by the equation:

$$6FeK_2.Fe(CN)_6 + KClO_3 + 3H_2SO_4$$
$$\rightarrow \underbrace{FeK.Fe(CN)_6}_{\text{potash blue}} + 3K_2SO_4 + KCl + 3H_2O$$

A test for completion of oxidation consists of placing a drop of the suspension on filter paper. A clear *halo* forms round the blue spot. A drop of ammonium (or potassium) thiocyanate solution is placed in contact with the halo. The formation of a reddish brown colour indicates complete oxidation. If the blue is not fully oxidized no colour is produced.

The subsequent filtration and washing of the blue are very slow processes due to the low specific gravity and very fine particle size of the pigment. Complete removal of soluble salts is very difficult, but important, because their presence results in a very hard pigment.

Drying must be carried out at a low temperature, preferably at 50° to 60°C (122° to 140°F), otherwise the pigment dries to a hard mass which is difficult to grind and to disperse in paint media.

Direct process of manufacture

A hot 10 percent solution of ferric chloride is added to a boiling 10 percent solution of potassium ferrocyanide in the presence of hydrochloric acid. Ferric potassium ferrocyanide (Potash blue) is precipitated according to the equation:

$$FeCl_3 + K_4Fe(CN)_6 \rightarrow \underbrace{FeK.Fe(CN)_6}_{\text{potash blue}} + 3KCl$$

This is followed by filtration, washing and drying at a low temperature in a similar manner to that used in the Indirect process.

The Direct process is not as widely used as the Indirect one. The latter is cheaper and gives a brighter product.

Ammonia-soda or bronze blues

These differ from the potash blues in that the ammonium radical replaces the potassium and that sodium ferrocyanide is used in the manufacture.

They are usually manufactured by the indirect process which follows that outlined above for Potash blue. A hot 10 percent solution of sodium ferrocyanide is added to a hot 10 percent solution of ferrous sulphate in the presence of ammonium sulphate:

$$FeSO_4 + Na_4Fe(CN)_6 + (NH_4)_2SO_4 \rightarrow Fe(NH_4)_2.Fe(CN)_6 + 2Na_2SO_4$$

The white precipitate is then oxidized with sodium chlorate in the presence of sulphuric acid:

$$6Fe(NH_4)_2.Fe(CN)_6 + NaClO_3 + 3H_2SO_4$$
$$\rightarrow \underbrace{6FeNH_4.Fe(CN)_6}_{\text{bronze blue}} + 3(NH_4)_2SO_4 + NaCl + 3H_2O$$

The pigment is then filtered, washed and dried at 50° to 60°C (122° to 140°F).

Properties of Prussian blues

In view of the difficulties involved in the drying and subsequent grinding and dispersion processes, considerable work has been done on the "flushing" of Prussian Blue. This has proved a difficult operation in view of the hydrophilic or water-attracting properties of the blue. It shows no tendency to migrate into the oil phase. However, by the use of cationic or certain non-ionic surface active agents it is possible to change the character of the pigment surface. By this means Prussian Blue has been made easier both to filter (as a result of flocculation) and to flush. In the latter case the blue is converted to the hydrophobic (water repelling) condition.

The terms *non-bronze* and *bronze* refer to the degree of metallic bronze exhibited by dry films after dispersion of the pigments in paint or ink media. For paintwork no bronze is required and the potash blues are therefore used. Bronze blues are used in printing inks where a degree of bronze is often desired.

The specific gravity of Prussian Blue is 1·97 and the oil absorption, which is related to particle size, varies from 70 to 160. The extreme fineness of the particles results in a high tinting strength, but the opacity in paint media is poor.

It possesses good lightfastness and is resistant to solvents and resins. In the earlier types of oil paint, Prussian Blue sometimes lost its colour on storage, due, presumably, to reduction to the colourless potassium ferrous ferrocyanide. On exposure of the paint film to the air, however, the original colour was restored by oxidation.

Prussian Blue is resistant to mineral and organic acids but is decomposed by alkalis. If this latter test is carried out in a test tube and the contents then acidified with mineral acid, Prussian Blue is reproduced. The reactions can be represented as follows:

$$FeK.Fe(CN)_6 + 3KOH \rightarrow K_4Fe(CN)_6 + Fe(OH)_3$$
$$K_4Fe(CN)_6 + Fe(OH)_3 + 3HCl \rightarrow FeK.Fe(CN)_6 + 3KCl + 3H_2O$$

The reactions with acid and alkali are, therefore, the opposite of those with Ultramarine Blue.

Prussian Blue is therefore unsuitable for use in paints to be applied to alkaline surfaces such as lime plasters and cements, or in water paints.

Uses

Non-bronze blues have been used extensively in the manufacture of Chrome or Brunswick greens. They were used widely in paints until the introduction of the phthalocyanine blues (Chapter 8), but in recent years their use has dropped considerably. Bronze blues are employed mainly in printing inks.

Identification tests

When heated strongly in an open crucible, Prussian Blue burns, emitting fumes of cyanogen (these are very poisonous and should not be inhaled) and leaves a residue consisting largely of ferric oxide.

It is unattacked by mineral acids but decomposed by alkali.

Other varieties of Prussian Blue

Brunswick Blue. A reduced blue produced either by precipitation of the Prussian Blue on to a base such as blanc fixe or barytes or by addition of the base to the Prussian Blue suspension before filtration. The blue is then more easily filtered, gives less trouble in drying, and is more readily dispersed. However, the presence of the base is often not desired.

Milori Blue. This term is sometimes applied to the best and palest potash blues which are completely devoid of bronze.

Chinese Blue. A Prussian Blue with a high bronze lustre.

Toxicity. Prussian Blue is considered to present a very low health hazard and this is supported by a long history of freedom from ill effects in usage.

Fire can create a serious hazard. When Prussian Blue burns the highly dangerous hydrocyanic (prussic) acid can be evolved.

Cobalt Blue (CI Pigment Blue 28)

Cobalt oxide forms solid solutions with some other metallic oxides. These are highly coloured and possess great resistance to heat, light and chemicals.

Cobalt Blue is sometimes known as Thénard's Blue and is a solid solution of cobalt and aluminium oxides. It is produced by calcination at 1000° to 1200°C (1832° to 2192°F) of mixtures of the oxides, hydroxides, or salts of the two metals.

Properties

Chemical analysis figures of Cobalt Blue correspond with the composition

15 percent CoO, 81 percent Al_2O_3, 4 percent H_2O. It is a very bright and clean colour and almost transparent when dispersed in paint media. It is valued for its colour but more particularly for its extreme stability. It is completely resistant to light, heat, acids and alkalis, but is expensive.

Uses

Cobalt Blue is used occasionally in paints but is too expensive and lacking in opacity for general use. It is used as a pigment for artists' paints.

Identification

A good pointer is provided by the extreme stability. It produces a fine blue borax bead, which serves to differentiate it from Phthalocyanine Blue.

Cerulean Blue

A solid solution of cobalt oxide, tin oxide and silica. It is paler and greener than Cobalt Blue and possesses the same stability as the latter. It is too expensive for general use.

Cobalt Violet

This term is applied to a range of purples and violets produced by calcination of cobalt phosphate alone or in admixture with other metal phosphates. They are extremely stable but are weak tinctorially.

GREEN PIGMENTS

For a number of years the green pigments in general use in the paint industry were confined to two groups of colours, both of which contained the metal chromium. These were (1) the chrome (or Brunswick) greens and (2) the oxide and hydrated oxide of chromium. The latter two have always been used for special purposes, but the chrome greens have been widely used in both decorative and industrial paints.

CHROME GREENS (CI Pigment Green 15)

The pigments are basically mixtures of lead chromes and Prussian Blue and, as mixtures, they exhibit the properties of the two components. They form a range of colours which have been standardized and are set out in BS 381C [1]. They are:

Grass green	British Standard Colour 318			
Brilliant green	,,	,,	,,	321
Light Brunswick green	,,	,,	,,	325
Mid Brunswick green	,,	,,	,,	326
Deep Brunswick green	,,	,,	,,	327

By the addition of certain other pigments such as raw sienna, red oxide and/or black oxide, a range of bronze greens is produced. These are generally deeper and redder than the corresponding chrome greens and are also set out in BS 381C. Their general properties are similar to those of the chrome greens.

Manufacture of chrome greens

It is generally agreed that the most satisfactory chrome greens are the so-called "struck" greens which are made by precipitating or "striking" the lead chrome in the presence of Prussian Blue. In this way the lead chrome is virtually precipitated on to the blue and yields a pigment from which the blue has less tendency to separate in the paint film (see Flooding and floating, *below*).

The greens can also be prepared by blending the dry pigments (there is a fire risk if they are dry ground) or by mixing aqueous suspensions of the two ingredients. In paint production, and particularly when non-standard colours are required, the dry chrome and blue are often dispersed in the paint medium. In all these cases, however, the blue has a greater tendency to separate than in the struck greens.

Properties and uses

The relative proportions of lead chrome and Prussian Blue can vary widely (the blue content can vary from 2 to 35 percent) and consequently figures for oil absorption and specific gravity can also show considerable variation.

Chrome greens are non-bleeding in paint media and exhibit good light-fastness, opacity, and durability when exposed in paint films. In industrial and urban atmospheres containing sulphur dioxide, surface "blueing" takes place as a result of reduction of the chromate radical. They are sufficiently heat-resistant to withstand normal stoving conditions, i.e. $\frac{1}{2}$ to 1 hour at 120°C (248°F).

The Prussian Blue ingredient is attacked by alkali, and consequently chrome greens cannot be used in paints for alkaline conditions.

They are subject to the restrictions of the Lead Paints Regulations and on account of toxicity they are no longer used in decorative or toy paints but continue to be used in certain industrial maintenance paints where the toxicity does not present a serious hazard. "Leadless" greens for use in decorative and similar paints are based on mixtures of organic yellows and phthalocyanine blue or green.

Flooding and floating

Depending on conditions of air circulation and humidity, paint films containing chrome greens can dry to a darker shade than the liquid paint due to an increase in Prussian Blue in the surface. This increase in concentration of the blue appears to be associated with the type of solvent evaporation and with convection currents in the fluid paint film.

If the deepening of surface colour is uniform the change is known as "flooding" and this occurs under conditions of "diffusive" solvent evaporation when disturbing influences such as draughts are absent.

More often, however, the surface has a mottled appearance and close examination will show the presence of Bénard cells. These are often hexagonal in shape and show differences in colour concentration between the centre and edges. These cells result from convection currents set up in the drying film, together with an "ebullioscopic" type of solvent evaporation from the surface. This occurs when air movement removes the solvent vapour and equilibrium conditions are not reached.

"Floating" can be converted to the uniform and less objectionable flooding by the addition of certain silicone oils at the rate of 1 part in about 40,000 parts of paint.

The introduction of a degree of thixotropy or structure by a thixotropic medium, hydrogenated castor oil, Bentone [2] or extenders such as Celite [3] or prepared silica [4] will reduce the film mobility and so reduce the intensity of colour change. A complete change of pigment is often the only satisfactory remedy. Flooding and floating can be exhibited by other mixtures of pigments which differ in specific gravity and surface characteristics. Greys based on titanium dioxide and carbon black can be particularly troublesome; blues made from mixtures of titanium dioxide and phthalocyanine blue can exhibit the defects, but this mixture responds more readily to remedial measures.

Floating can occur with some single pigments, for example, with toluidine red in a long-oil alkyd medium, but the effect is not usually noticeable.

Identification tests. As mixtures of lead chromes and Prussian blue, chrome greens show the reactions of both these pigments.

Toxicity of chrome greens. The toxic hazards are largely those associated with the lead chromes (page 83). The toxicity of Prussian Blue is very low (page 102).

Chromium oxide (CI Pigment Green 17)

This consists, as the name suggests, of anhydrous chromium oxide, Cr_2O_3.

Manufacture

Several methods have been described in the literature among which are the following:

(i) A mixture of potassium dichromate and sulphur is ignited:

$$K_2Cr_2O_7 + S \rightarrow Cr_2O_3 + K_2SO_4$$

The product is ground in water, filtered, washed, dried and reground.

(ii) Sodium dichromate is heated to 300°C in the presence of hydrogen:

$$Na_2Cr_2O_7 + 3H_2 \rightarrow Cr_2O_3 + Na_2O + 3H_2O$$

The fused mass is ground in water, filtered, washed, dried and reground.

Properties and uses

Chromium oxide is a sage-green colour and is dull and chalky compared with the chrome greens. It possesses the following characteristics:

Specific gravity	5·2
Oil absorption	10

The opacity is very high as is also the resistance to light, heat, acids and alkalis. It is extremely durable and is used under any conditions where resistance is more important than a bright colour. It can safely be used in water paints and distempers and finds wide use in colouring concrete and roof tiles, plastics; printing inks, and ceramics.

Identification tests

Chromium oxide is characterized by general inertness and colour. Fusion with a mixture of anhydrous sodium carbonate (5 parts) and sodium peroxide (1 part) converts it to sodium chromate which can be extracted with water. Acidification followed by addition of hydrogen peroxide gives the fugitive blue colour which is characteristic of chromate and hence of chromium.

Toxicity. Chromium oxide presents a very low toxic risk.

Guignet's Green or Viridian (CI Pigment Green 18)

This is a hydrated oxide of chromium and is generally assigned the formula $Cr_2O_3 . 2H_2O$.

Manufacture

Several methods have been described using sodium dichromate as the starting material. In one of these it is fused with boric acid when the following reaction takes place:

$$Na_2Cr_2O_7 + 16H_3BO_3 \rightarrow Cr_2(B_4O_7)_3 + 24H_2O + Na_2B_4O_7 + 3[O]$$

The fused mass is then extracted with water when hydrolysis of the chromium pyroborate yields the hydrated chromium oxide:

$$Cr_2(B_4O_7)_3 + 20H_2O \rightarrow Cr_2O_3 . 2H_2O + 12H_3BO_3$$

The pigment is filtered, washed well, dried at 65° to 70°C (149° to 158°F) and ground.

In another process the sodium dichromate is fused with ammonium phosphate in place of boric acid. The mass is hydrolysed and processed as before.

Properties and uses

Guignet's Green is a much brighter pigment than chromium oxide and possesses the following physical characteristics:

Specific gravity	3·2
Oil absorption	63

It has very poor opacity but is fast to light, alkali and all types of paint media. The resistance to acids and heat is not as high as that of chromium oxide. It is unaffected by cold mineral acids but dissolves in hot concentrated hydrochrloric acid. On strong heating it is decomposed at about 250°C (482°F). Lack of opacity and rather high cost limits its use in paints, but it is a satisfactory stainer for water paints and distempers. However, it is gradually being replaced in these paints by organic pigments.

Identification tests

It is turned black on very strong heating. It is soluble in hot concentrated hydrochloric acid to give a solution of chromium chloride. This gives the normal reactions of chromium ions. Fusion with a mixture of anhydrous sodium carbonate and sodium peroxide gives sodium chromate (see Chromium oxide).

References to Chapter 6

[1] BS 381C:1964 Colours for specific purposes. British Standards Institution, 2 Park Street, London.
[2] Steetley Chemicals (NL Industries Ltd).
[3] Johns-Manville, Denver, Colorado.
[4] "Aerosil" (Degussa); "Gasil" (Joseph Crosfield).

7 Black, metallic, and miscellaneous pigments

BLACK PIGMENTS

The bulk of the black pigments used in the modern paint industry consist of amorphous carbon black either in a fairly pure state (carbon and vegetable blacks) or in admixture with inorganic matter (bone blacks). Other blacks in wide use comprise graphite and the black oxides of iron.

CARBON BLACKS (CI Pigment Black 7)

These are derived from petroleum gases and liquids and are classed as Channel or Furnace blacks according to the method of production.

Manufacture — channel process

The basis of the process is the burning of hydrocarbon gases in specially designed burners under conditions of limited air supply so that combustion is incomplete. The smoky flames produced impinge on lengths of channel iron on which the carbon is deposited and from which it is removed by mechanical scrapers. The carbon black is blown through screens to remove any grit derived from the channels.

The quality of the black can be controlled by the shape of the flame, distance from channel surface, amount of air, and quality of the gas.

When first produced, channel black is very fluffy due to entrained air and gaseous by-products resulting from incomplete combustion. It is therefore compressed or converted to beads by either wet or dry pelletizing.

Wet pelletizing. The carbon black is passed into a cylinder fitted with a central rotating shaft to which spokes are attached. Water is added at the same time and the carbon black forms pellets which are then dried.

Dry pelletizing. The carbon black particles are caused to adhere together by passing the black slowly through rotating drums.

The efficiency of the channel process is very low and prevention of smoke emission into the atmosphere has been found very difficult. For these reasons and the fact that considerable improvements have been effected in the furnace process, furnace blacks are being used to replace channel blacks in a number of applications.

Gas black process [1]

The basic principle is similar to that of the channel process in that the petroleum product is burned in a limited supply of air. The two processes differ, however, in two respects, (a) the flames impinge on cooled rollers from which the carbon black is removed by scraper, and (b) oil is used in place of gas, giving the gas black process greater flexibility. Subsequent processing of the carbon black is on similar lines to the channel process. Carbon blacks made by this process are sometimes known as "impingement" blacks.

Furnace process

The older furnace blacks were produced by combustion of gases in a limited supply of air in a refractory-lined furnace. A single large flame was employed and the products, containing the carbon black, were quenched with water before passing to bag filters or electrostatic precipitators. The resulting carbon black was used only in the rubber industry.

The modern process uses a similar type of furnace but employs oil instead of gas (although mixtures are sometimes used) and by varying the conditions a wide range of carbon blacks can be produced, covering the grades required for paints and printing inks as well as rubber. A high proportion of the carbon black consumed at the present time is made by the furnace process.

Oxidized blacks carry carboxyl and other groups on the surface and show improvements in certain characteristics, notably flow properties and ease of dispersion.

Other processes producing carbon black use thermal decomposition rather than incomplete combustion. They are the Thermal, Electric Arc and Acetylene Black processes, but their output is considerably less than that of the Furnace and Channel processes.

Properties of carbon blacks

The physical characteristics of the grades of carbon black used in paints and inks are summarized in Table 7.1 [1].

Carbon blacks are classified according to the intensity of blackness which is quoted on the Nigrometer scale (Table 7.1) on which lower figures indicate more intense blackness. The intensity of blackness increases with decreasing particle size.

Carbon blacks are hygroscopic and, if exposed to a moist atmosphere, can absorb up to 15 percent of moisture, in consequence of which they are difficult to disperse in organic media. The oxidized types are an exception.

As a result of the almost total absorption of incident light and the fine particle size, carbon blacks have very high opacity. In a brushing, air-drying paint, 25–35 grams of carbon black per litre is generally sufficient to give a completely opaque film.

Table 7.1

Black	Mean particle size range	Blackness (Nigrometer)	Surface area BET* m²/gram	Oil absorption	pH of aqueous suspension
Channel	5–30 mμ	58– 85	100–950	80–150	4– 8
Gas black	10–30 mμ	58– 85	90–800	80–150	4– 8
Furnace	10–80 mμ	80–105	20–550	80–150	8–10

Notes (a) $m\mu = $ millimicron $= 10^{-9}$ metre.
      (b) Blacks in the 20–30 mμ range are suitable for rubber.
      (c) Each type of black is supplied in a number of grades (degrees of blackness) the particle sizes and surface areas of which fall within the ranges quoted.
      (d) Figures for surface area by other methods, e.g. from average particle diameter, are often appreciably lower than the BET figures owing to the "porous" nature of carbon black "particles". These particles appear to consist of agglomerates of much smaller units, giving the particles a "porous" structure. The ultimate units appear to have a distorted crystalline structure akin to graphite [2], [3].

They are completely fast to light and chemicals (with the exception of concentrated nitric acid) and to the temperatures which can be tolerated by most paint systems.

When used in paints, carbon blacks tend to adsorb materials from the medium. Thus an air-drying black paint gradually loses its drying properties on storage (unless the paint contains an additive, e.g. Nuact [4] to prevent this) as a result of the adsorption of part of the drier metals by the carbon black. The drying properties are recovered to some extent by re-milling the paint. On prolonged storage, black air-drying paints also show an increase in viscosity due to adsorption of lower molecular weight fractions of the polymer.

The fine particle size of carbon blacks (Fig. 7.1) is accompanied by high tinting strength, but many blacks, when mixed with white, produce greys with a brownish tint. Cleaner, bluer tints are produced by the lamp blacks (*below*) which have generally been preferred for this purpose. Some of the newer carbon blacks also give a bluish grey on admixture with white.

Uses of carbon blacks

Carbon blacks are used in conventional paint systems and in inks. Care should be taken, however, with some chemically-cured systems to ensure that the pigment does not impair the curing reaction.

They are marketed in several physical forms ("fluffy", dense, beads) and also dispersed in "chips" (in plasticized cellulose nitrate), in plasticizers, and in stainers for organic and aqueous paints.

Identification. They are inert to all solvents and acids (with the exception of concentrated nitric acid when vigorous oxidation can ensue) and alkalis. On ignition they burn, leaving no appreciable ash. They are differentiated from lamp black by the more intense self colour.

---

* Brunauer, Emmett & Teller method.

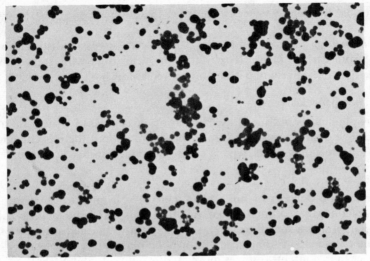

Courtesy: Paint Research Association

Fig. 7.1   Photomicrograph of carbon black (× 48 000)

Lamp (Vegetable) black

The term "vegetable black" was originally given to the product obtained by burning vegetable oils in a limited supply of air. It was marketed in two grades, known as vegetable and lamp blacks, but the two terms are now used for the one product.

Manufacture. The present-day method consists in incomplete combustion of liquid petroleum products contained in a cast-iron pan situated under a flue. The lamp black is collected by cyclone and filters. The particle size can be varied by varying the conditions of combustion.

Properties

Lamp black has the following physical characteristics —

Specific gravity 1·8–1·9   BET surface area 20–50 m^2/gram
Oil absorption 90–150   pH of aqueous dispersions 6–9
Nigrometer value 93–102   Mean particle size range 50–120 mμ

It is coarser than carbon black and the colour is duller and greyer. When mixed with whites it gives clean bluish greys and is used as a tinting pigment rather than as a self colour.

Lamp black is fast to light, paint media, chemicals (with the exception of concentrated nitric acid when vigorous oxidation can take place) and to the temperatures which can be tolerated by most paint systems. It can contain up to 2 percent of oily matter and is therefore difficult to disperse in water but fairly readily dispersed in organic media.

Uses. Lamp black is used as a stainer in paints. It is also used in rubber and inks.

Identification. The dull and grey self colour serves to distinguish lamp black from carbon black. It is wetted by water only with great difficulty, and on ignition it burns and leaves no appreciable ash.

Drop (Bone) black

This pigment is produced by calcination of degreased bones and animal refuse. The quality is assessed on carbon content and there is considerable variation depending on the type of bone and conditions of calcination.

Properties

The carbon content of bone black can vary from 5 to 20 percent, the balance consisting principally of calcium phosphate and a little calcium carbonate. It has a specific gravity of 2·6 to 2·8 and an oil absorption of 30 to 40. The colour and tinting strength of bone black are variable and depend on the carbon content. It mixes more readily with oils and water than do the carbon blacks but it is similar to carbon black in its adsorption properties. Thus in paint it tends to adsorb driers from the medium and it is sometimes used in water to adsorb colouring-matters. Frequently a special grade, known as "animal charcoal", is made for this purpose.

Uses. Bone black is used in fillers and in the cheaper types of black undercoat.

Identification. On ignition the carbon burns off leaving a white or slightly pink residue. This residue is soluble in hydrochloric or nitric acid and the solution gives the normal reactions for calcium and phosphate ions.

Graphite

This material is a crystalline form of carbon and is obtained from natural sources (chiefly from Sri Lanka) or is manufactured from a high-carbon type coal. It is also known as "black lead" or "plumbago". The particles are in the form of hexagonal plates.

Natural graphite is ground, washed with dilute acid, then with water, and dried. The quality is assessed on the carbon content and is very variable.

Manufactured graphite is prepared from anthracite by the Acheson process. The anthracite is finely powdered and packed tightly into crucibles. These are then heated to 1200° to 1500°C (2192° to 2732°F) in absence of air and in an electric furnace. There is a 98 percent conversion to graphite. The success of the process depends on the availability of a cheap source of electricity and it is therefore worked extensively in Canada.

Properties

Graphite is a dark grey material with a metallic lustre and a characteristic slippery feel. The specific gravity varies, the natural being about 2·5 and the manufactured material 2·25.

It has good opacity and the plate-like structure and characteristic surface give it great spreading power. When made into paint the particles leaf with their planes parallel to the surface and overlapping. The film then possesses high resistance to moisture penetration. Graphite also has a tendency to *grip* metal surfaces, resulting in good adhesion.

Uses

Graphite is used in maintenance paints for structural steelwork and also in corrosion-inhibiting primers. In the latter case it is often mixed with red lead. Graphite is also used in lubricants and pencils.

Identification

Graphite has a characteristic metallic appearance and slippery feel. On ignition the carbon burns and leaves, with natural material, a white siliceous ash. With manufactured graphite, very little ash is obtained.

IRON OXIDE BLACKS

Natural black oxide of iron

This material is also known as Magnetite or Black Magnetic Oxide and has the formula Fe_3O_4. It is very coarse and dense and is used in fillers and stoppers but not in paints.

Micaceous iron oxide

This is a unique form of iron oxide the particles of which are glistening flakes resembling mica. It is sometimes known as "specular haematite ore". The pigment as supplied is a greyish or greenish black, dense and coarse product, the average particle diameter of which can be as great as 60–70 μm. Finer grades are available and have to be used when the paints are applied by airless spray. The lamellar form of the particles is destroyed by grinding, resulting in the formation of red iron oxide.

Micaceous iron oxide contains 85 percent and upwards of iron oxide (expressed as Fe_2O_3). The specific gravity is 4·7 and the oil absorption 20–25. It is insoluble in all paint media and resistant to most chemicals.

Uses

Micaceous iron oxide is used in maintenance paints for steel structures where protection rather than decoration is the important consideration. It is unaffected by weathering and forms a moisture-resistant barrier in a paint film by the leafing properties of the lamellar particles. The chemical inertness enables it to be used in a variety of media among which tung oil/phenolic, chlorinated rubber, and epoxies are the most commonly used.

Identification

The characteristic appearance is often sufficient. Chemical identification as an iron oxide involves solution in concentrated hydrochloric acid followed by dilution and test for ferric ions.

Precipitated black oxide, Fe_3O_4

The manufacture of this pigment is on similar lines to that of the yellow and red oxides (Chapter 5). A solution of ferrous sulphate is treated with alkali to precipitate ferrous hydroxide. This is then oxidized to the black oxide by blowing air through the suspension in open tanks:

$$6Fe(OH)_2 + O_2 \rightarrow 2Fe_3O_4 + 6H_2O$$

The conditions of oxidation are very carefully controlled. The pigment is then filtered, washed well, and dried.

Properties and uses

Precipitated black oxide of iron has a specific gravity of 4·7 and an oil absorption of 28 to 30. It has a fine soft texture and moderate tinting strength. It is used in undercoats where carbon black can lead to increase in viscosity, and sometimes in oil-type finishes.

The lightfastness is good as is also the fastness to solvents and paint media generally. The heat resistance is satisfactory for normal stoving conditions, i.e. 120°C (248°F) for half an hour, but exposure to air for prolonged periods at high temperatures leads to some reddening of the colour. It is fast to alkalis but is soluble in hot concentrated hydrochloric acid.

Identification

Ignition in air results in oxidation to red Fe_2O_3:

$$2Fe_3O_4 + \tfrac{1}{2}O_2 \rightarrow 3Fe_2O_3$$

It is soluble in hot concentrated hydrochloric acid and the solution gives the normal reactions for ferric ions.

METALLIC PIGMENTS

ALUMINIUM (CI Pigment Metal 1)

Finely divided aluminium powder appeared at about the beginning of this century and its use as a paint pigment developed gradually from that time. Considerable impetus was given to its use by the introduction of the paste pigment, so avoiding the dust nuisance which was, and is, a feature of the powder. Today both types are in use, but the consumption of powder in paint is very small compared with that of the paste.

Aluminium pigment is an interesting material in that the particles are in the form of small flakes which possess a high metallic lustre and exhibit the property of leafing in organic media. The particles orientate themselves with their planes parallel to the surface of the paint film.

Manufacture

The early method consisted of beating the powdered metal into flakes by multiple hammer mills in the presence of a suitable lubricant, the most widely used being stearic acid.

Modern methods employ ball mills in which the flattening of the metal particles into flakes is effected by the cascading action of the steel balls. The mill is charged with the powdered aluminium, stearic acid and white spirit. Quality of product and output are determined by the size of charge of steel balls, charge of aluminium and white spirit, and speed of rotation of the mill. The pigment slurry from the mill is filter-pressed and the press cakes, containing 66 to 68 percent of metal, comprise the aluminium paste of commerce. If powder is required, the white spirit is removed under vacuum at as low a temperature as is practicable.

Types of aluminium pigment

The pigment is marketed in both paste and powder forms, and each of these is available in leafing or non-leafing types.

Aluminium paste

The leafing and non-leafing types are graded according to particle size into the following grades:

Standard. A general purpose grade, very widely used, and containing approximately 67 percent pigment.

Polished or Brilliant (leafing type only). This is designed to give a more mirror-like finish than the standard. The paste usually contains about 75 percent pigment.

Naphtha base Standard. A paste containing 67 percent aluminium in which the solvent is naphtha in place of white spirit. It is useful when the paint medium is incompatible with white spirit.

Ink and Fine Lining. These comprise the finest particle-size pigments.

Aluminium powder

This is also marketed in leafing and non-leafing types and in various grades based on particle size.

Properties and uses

Paste, Leafing Type. The outstanding property of this type is the ability of the particles to leaf in organic media. The flakes lie close to the surface, and overlapping results in an almost continuous bright metallic film. The leafing properties depend on the presence of a thin layer of stearic acid on the surface of the particles and are destroyed by the presence of lead driers in the paint.

In spite of the presence of a layer of stearic acid on the surface of the particles, aluminium pigment is very reactive and will react with organic acids present in the paint medium. This can result in loss of leafing power and brilliance, but hydrogen can also be liberated, resulting in pressure development in the tins. The acid value of the medium used should therefore be as low as possible and should not, in any event, exceed 4.

As is well known, aluminium is amphoteric and so is attacked by both acid and alkali. The presence of the stearic acid causes an "induction period" which is followed by a vigorous reaction.

Aluminium is non-toxic and the leafing paste is widely used in decorative and metal priming paints and in industrial finishes. In conjunction with silicone resins it is used in paints for high-temperature resistance. It is also used in bituminous aluminium paints where it increases the light — and heat — reflectance of the surface.

Paste, Non-Leafing Type. There are certain paints where the continuous metallic lustre is not required but where the presence of the flake particles is desirable. These include primers for woodwork and the industrial finishes where special effects are required, for example, polychromatic and hammer finishes. The non-leafing paste is preferred for these materials.

Powder, Leafing. The properties are generally similar to those of the paste, but the finishes produced are somewhat brighter but less smooth. For this reason, and the dust nuisance, the powders are now used to a very limited extent.

Powder, Non-Leafing. The uses of this material are similar to those of the paste type, but the latter is used whenever possible to avoid the dust hazard.

Identification

Aluminium pigment has a characteristic appearance and when heated strongly in the air will burn to the oxide. It is soluble in acid and alkali. The acid solution gives the normal reactions for aluminium ions.

Lead

Metallic lead is produced in a very fine state of division, but on account of the great affinity for oxygen, it is supplied as a dispersion in water or in an organic medium. The finely divided metal is strongly hydrophobic and so flushing from the aqueous suspension into an organic medium presents no difficulties.

Properties and uses

It is a very finely divided pigment with a mean particle size of less than one micrometre. Its main use is in corrosion-inhibiting primers for steelwork and here it has been successfully used in a number of media of different types including linseed oil, isomerized rubber and epoxy resins.

The mechanism by which lead protects iron has created a great deal of interest. Electrode potential measurements on the metals in the massive state indicate that metallic lead should be cathodic to steel and when in contact should stimulate corrosion. However, Whitby [5] has shown that in a film of metallic lead primer, the lead is anodic to steel and could therefore protect by sacrificial corrosion. This mechanism would require metal contact and this has been found to be unnecessary for the effective functioning of metallic lead primers. Satisfactory results have been obtained when the metallic lead content of the pigment has been as low as 50 percent.

According to Wild [6], metallic lead pigment absorbs oxygen very readily from its environment and therefore probably protects by inhibiting the cathodic reaction. This "oxygen barrier" idea would explain the efficiency of metallic lead in media other than drying oils such as isomerized rubber, chlorinated rubber, and epoxy resins.

In drying oils and other oxidizing media it is possible that some passivation of the steel surface takes place as a result of the formation of lead soaps of scission products, as suggested by Mayne and van Rooyen [7] for red lead.

Metallic lead is a highly toxic pigment and its use is subject to the Lead Paints Regulations. Application of primers containing metallic lead is by brush or hand roller only, but on account of its high toxicity the use of the pigment is declining.

Identification. Metallic lead is dissolved by nitric acid and the solution gives the normal reactions for lead ions.

Stainless steel flake

This pigment is available in both dry and paste forms. When used in paints, the films possess exceptional hardness and very good corrosion-resisting properties. It has high opacity and is used in situations where corrosive or chemical attack is likely to be severe.

Bronze powders

This term covers a range of copper alloys some of which are akin to the brasses. The paler and brighter shades are marketed under the name "gold powder".

They are manufactured in a manner similar to that for aluminium and the particles are in the form of tiny flakes. Their specific gravity is very considerably higher than that of aluminium and although they possess leafing properties these are less pronounced.

In general the bronzes are very sensitive to small amounts of acidity in the medium and tarnish rapidly on exterior exposure. They are used therefore for interior decoration and for this purpose are often dispersed in straight polymer solutions such as polyvinyl acetate or polystyrene.

Identification tests. The appearance of these pigments is characteristic. When warmed with 50 percent nitric acid (this should be done very cautiously) bronze powder dissolves. The solution gives the reactions of copper and (usually) zinc ions.

ZINC

This is marketed as a blue-grey powder under the name *Zinc Dust* containing 95 to 97 percent metallic zinc. The balance is zinc oxide formed by oxidation of the finely divided metal.

Manufacture

Metallic zinc (spelter) is vaporized and the vapour is led into large condensers. Under the conditions prevailing, the metal condenses as a dust which is removed by suitable methods. By rigid control of conditions the particle size distribution can be controlled to within fairly narrow limits.

Properties

Zinc dust is supplied in two grades known as Standard and Superfine. These compare as follows:

	Standard	Superfine
Specific gravity	7·1	7·1
Average particle size range	5–9 μm	2·25–3·0 μm

The particles are spherical in shape. The pigment has very good opacity but very low tinting strength. It is very reactive.

Uses of zinc dust

In the paint industry the use of zinc dust is largely confined to the protection of steel surfaces. Primers for steelwork contain between 92 and 95 percent metallic zinc in the dry film and are therefore designated *zinc rich*. The following types of binder have been employed satisfactorily: polystyrene, isomerized and chlorinated rubbers, epoxies, acrylics, and vinyls.

The mechanism by which zinc-rich primers protect steel is electrochemical and is akin to the protection afforded by zinc in galvanized iron. For this to be effective it is essential that there be actual contact between the steel surface and the metallic zinc. The surface of the steel must therefore be perfectly clean. There must also be contact between the particles of metallic zinc in the film, so that an adequate supply of metal is available. If therefore the metal content of the zinc-rich film falls below 92 percent, sufficient binder is present to isolate the particles. If the metal content is above 95 percent insufficient binder is present to provide the necessary adhesion.

When iron and zinc are in contact in a corrosive environment, the zinc becomes anodic to the iron and is preferentially corroded. The action is therefore described as *sacrificial protection*.

In the United States and to some extent in this country, zinc dust (80 parts) is mixed with zinc oxide (20 parts) in drying oil or oleoresinous media to produce corrosion-inhibiting primers for general use on steel. The performance of these paints is of a high order but the mechanism of action has not been completely elucidated.

Inorganic zinc silicate primers

Zinc dust will react with silicates of alkali metals to produce films which are hard and impervious. The reaction with sodium and, later, lithium silicate was used for many years to protect steelwork under severely corrosive conditions. The film consisted of zinc silicate together with unreacted zinc which provided sacrificial protection. The grade of alkali silicate used was important in order to minimize formation of the water-soluble sodium or lithium zincate. These paints were supplied as two-pack systems, the zinc and silicate solutions being packed separately and mixed immediately before use.

More recently the alkali metal silicates have been replaced by a solution of a pre-hydrolysed silicon ester. The systems are two-pack and require moisture to cure to a hard, adherent and impervious film of zinc silicate containing metallic zinc. Such primers give excellent protection to steel under severe conditions, provided the steel has been thoroughly cleaned. It is generally blast-cleaned to SA 3 of the Swedish Standard [8] or first quality to BS 4232 [9].

Identification tests

On heating in air, zinc dust burns to form the oxide (yellow when hot, white when cold). It reacts vigorously with dilute hydrochloric acid to give a solution exhibiting the normal reactions of zinc ions.

Toxicity. Metallic zinc presents no significant toxic hazard, but the dust is classified as a "nuisance particulate". Zinc oxide is formed by welding or flame-cutting through zinc-coated metal and this can cause "zinc fume fever". The effects are unpleasant but transitory. TLV zinc oxide fume 5 mg/m^3.

MISCELLANEOUS PIGMENTS

Calcium plumbate

Calcium plumbate differs from the other pigments containing lead, with the single exception of red lead, in that the lead occurs in the anion. The composition is closely related to that of red lead, as can be seen when the two are compared:

Calcium plumbate	$2CaO . PbO_2$
Red lead	$2PbO . PbO_2$

A calcium/lead plumbate, $CaO . PbO . PbO_2$, also exists.

Manufacture

Commercial calcium hydroxide and litharge are heated together in the proportions of two molecular weights of the former and one molecular weight of the latter. The optimum temperature is about 700°C (1292°F) and the time of heating 8 to 10 hours.

Properties and uses

The calcium plumbate of commerce is a cream or light buff-coloured soft powder. The colour varies with the purity of the raw materials, and the use of pure lime will give a white product. The colour can also be influenced by slight variations in the ratio of lime to litharge.

The specific gravity is 5·7 and oil absorption 10 to 17. It has high opacity and is readily dispersed in oil media on account of its alkaline reaction. Cold water extraction yields about 0·5 percent soluble matter, essentially calcium hydroxide. Using the British Standard method [10], the soluble matter is much higher and variable (2 to 4 percent). It is thought that the alkaline nature contributes to the rust-inhibiting properties of the pigment.

Calcium plumbate is used extensively in corrosion-inhibiting primers for steelwork and can be employed in a number of media such as linseed oils, stand oils, long oil alkyds and chlorinated rubber. The mechanism whereby calcium plumbate protects steel is discussed by Read [11], who considers it probable that an inhibitive film is laid down at both anodic and cathodic areas on the steel. This is supported by Evans [12] who has described rust inhibition by some calcium compounds.

The priming of new galvanized iron has, in the past, been preceded by some form of surface treatment in view of the poor adhesion between the new metal and the primer. Calcium plumbate primers adhere satisfactorily to the untreated metal and are now used widely for this purpose. They have also given promising results as primers for woodwork and hardboard.

Identification

The aqueous extract shows an alkaline reaction. Treatment with dilute nitric acid yields a solution of calcium ions and a brown residue of PbO_2. This is filtered off, redispersed in dilute nitric acid, and a few crystals of sodium nitrite or oxalic acid added. The lead dioxide is reduced to the monoxide, which dissolves to a colourless solution giving the normal reactions for lead ions.

Toxicity

As lead compounds, the manufacture and use of calcium plumbates are controlled by the Lead Paint Regulations. By the same token it is probable that, in due course, calcium plumbate will be phased out.

Barium metaborate

The pure compound, $Ba(BO_2)_2$, is appreciably soluble in water but the solubility can be reduced to 0·3 to 0·4 grams per 100 cm^3 by precipitation as a double salt with silicate. This "modified" form, to which the formula $3BaO.3B_2O_3.2SiO_2$ has been assigned [13], is the type usually marketed. It has a specific gravity of 3·25 to 3·35 and an alkaline (pH 9·5) aqueous extract. Barium metaborate has been recommended for a variety of uses including corrosion inhibition, fire retardance, mould and bacteria control, and chalk resistance.

LUMINOUS PIGMENTS

Generally the colour of material objects is the result of selective absorption and reflection of wavelengths in the visible spectrum (Chapter 2). Certain substances, however, possess, in addition, the property of absorbing in the invisible ultraviolet waveband and re-emitting part of the energy at longer wavelengths within the visible spectrum. Such materials may be of two types:

Fluorescent pigments. The luminosity exists only as long as the substance is exposed to the exciting radiation and disappears almost immediately on its removal.

Phosphorescent pigments. These store energy from the exciting radiation so that light emission persists after this is removed. These pigments possess an after-glow which persists for short or long periods depending on their composition.

Fluorescent pigments

A widely used type contains fluorescent dyestuffs in a hard resin base. They are thermoplastic and possess softening points between 100°C and 120°C (212° and 248°F).

They are daylight-activated and are used in high-visibility paints, but as they are translucent the paints are applied over a flat white ground coat.

The performance of the paints can be improved by a coat of protective lacquer (PVA, acrylic, or cellulose) but the exterior durability is limited to about 12 to 18 months.

An inorganic type of fluorescent pigment is used in the coatings applied to the interior surfaces of fluorescent electric lamps. Among the materials used are calcium and magnesium tungstates, zinc orthosilicate and cadmium borate.

Phosphorescent pigments

This class includes the materials commonly known as luminous pigments. They are, in the main, sulphides of metals, of which the most important are zinc, calcium, strontium, barium, and cadmium.

The phosphorescence is obtained by preparation of the compounds in the form of *activated* crystals, the activity being produced by minute traces of compounds of copper and manganese; some of the rare earth metals are also useful in improving the luminosity. These activating materials are known as phosphorogens and appear to exist as solid solutions in the crystals of the parent sulphide.

In the manufacture of luminous pigments scrupulous care has to be taken to exclude traces of materials which can destroy the luminosity. Chief among these are compounds of lead, iron and chromium.

The process of manufacture depends on the preparation of the sulphide in a pure state. This is dried and converted to the "phosphor" or active material by calcination with the activating agent, a flux, and a reducing agent. The flux comprises a eutectic mixture of salts of sodium or potassium, e.g. borax, sodium chloride, sodium sulphate or potassium sulphate, all of which must be pure.

The temperature of calcination varies with the nature of the metal and is generally between 800°C and 1100°C (1472° and 2012°F). After calcination the powder is disintegrated only, care being taken to avoid contact with any "poisonous" metals. Fine grinding reduces the luminosity, and the particle size range is from 50 to 150 μm.

The colour emitted by the luminous pigment depends on the composition and on the nature of the activating phosphorogen. The individual sulphides give the following colours:

Zinc sulphide	Yellowish-green
Cadmium sulphide	Red
Calcium sulphide	Violet
Strontium sulphide	Blue/Green
Barium sulphide	Yellow

Variations can be obtained by combinations of the individual pigments.

Moisture destroys the luminosity and they should, therefore, be stored under dry conditions.

Paints produced from these pigments contain the minimum amount of binder and have limited durability.

References to Chapter 7

[1] Degussa publication, "What is Carbon Black?"
[2] HESS, W. M. & GARRETT, M. D., The dispersion of carbon blacks in paint and ink systems. *JOCCA*, 1971, **54**, 24.
[3] DOLLIMORE, D., The characterisation of carbon black surfaces. *JOCCA*, 1971, **54**, 616.
[4] "Nuact" (Durham Raw Materials).
[5] Paint Research Association Technical Paper No. 96.
[6] WILD, G. L. E., Metallic lead pigment for anti-corrosive protection. *JOCCA*, 1965, **48**, 43.
[7] MAYNE, J. E. O. & VAN ROOYEN, *J. Appl. Chem.*, 1954, **4**, 384.
[8] SIS 05 59 00:1967. Swedish Standards Institution.

[9]   BS 4232:1967. Surface finish of blast-cleaned steel for painting. British Standards Institution.

[10] BS 3483:1974. Methods of testing pigments for paints. 24 parts. British Standards Institution.

[11] READ, N. J., Calcium plumbate for priming paints. *JOCCA*, 1958, **41**, 352.

[12] EVANS, U. R., *An Introduction to Metallic Corrosion*, 3rd edn. Edward Arnold, London, 1979.

[13] KHILIL, M. A., EL-SAAWY, S. M. & GHANEM, N. A., Laboratory preparation, modification and evaluation of inhibitive barium metaborate pigments. *Pig. & Resin Tech.*, 1981, **10**, No. 6, 4.

Bibliography

DANE, C. D., New light on fluorescent pigments. *Chem. Brit.*, 1977, **13**, No. 9, 335.

8 Organic pigments

The organic pigments offer a wider range and a greater variety of shades than do the inorganics. Although some are inferior in opacity, they possess good staining power, softness of texture and brilliance. A number of brilliant dyes and pigments can be obtained from plant or animal sources, but these have now been almost entirely superseded by the synthetic types. However, a few continue to be used for special purposes and are listed, together with others of historic interest, in Appendix E.

Sources

The aromatic hydrocarbons, derived largely from the petroleum industry, constitute the raw materials of a large proportion of the organic pigments. The hydrocarbons are converted to the dyes and pigments in stages involving the formation of the following classes of compounds.

Primaries. These are the first-stage products derived from the hydro-carbons. Examples are nitrobenzene, toluene sulphonic acid, and nitro-naphthalene.

Intermediates are derived from the primaries by further reaction which can involve more than one stage, e.g. aniline from nitrobenzene, aceto-acetanilide from aniline and acetoacetic ester.

Intermediates are converted directly to pigments or dyestuffs and the synthesis of these can be represented thus:

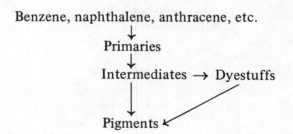

Colour-forming groups

The colour of organic pigments depends on the presence of certain groups in the molecule. These are known as *chromophores* and the com-monest are the following:

$$\rangle C = C \langle, \quad \rangle C = O, \quad \rangle C = S, \quad \rangle C = NH, \quad \rangle C = N-,$$

$$-N=N-, \quad -N=O, \quad -NO_2.$$

Some materials contain, in addition, salt-forming groups known as *auxochromes*. These are important in pigment manufacture. The principal salt-forming groups are:

$$\underbrace{-OH, \quad -SO_3H, \quad -COOH}_{\text{acidic}} \qquad \underbrace{-NH_2, \quad -NHR, \quad -NR_2}_{\text{basic}}$$

The molecule without the auxochrome, but containing the chromophore, is known as the *chromogen*.

General classification of organic pigments

It is possible to use a very broad classification and to divide organic pigments into pigment dyestuffs, toners, and lakes.

Pigment dyestuffs contain no salt-forming groups and comprise a wide variety of chemical types. These include azo compounds, metal complexes and polycyclic compounds.

Toners are produced by precipitation of acid dyestuffs by treatment with salts of certain heavy metals (mainly calcium, barium and manganese) or of basic dyestuffs with acidic reagents. Examples are Lithol toners and "Fanal" type pigments. The term "toner" is sometimes applied to pure pigment dyestuffs.

Lakes are "reduced" pigments prepared by precipitation on to white bases such as alumina (aluminium hydroxide), blanc fixe, china clay, etc. In some cases a reactive base such as alumina forms a co-ordination complex with the dyestuff and becomes an integral part of the pigment. Madder lake is of this type.

CHEMICAL CLASSIFICATION OF ORGANIC PIGMENTS

The great number of organic pigments known fall into a comparatively small number of chemical groups. Classification is by chromophore (as in the nitroso and azo groups) or by general structure (as in the anthraquinones). The following groups are the most widely known, but the number of examples quoted is but a small fraction of the whole range available.

Nitroso group

This group contains only one member of interest to the paint industry.

Pigment Green B (CI Pigment Green 8) is an interesting iron co-ordination compound produced from the bisulphite compound of a-nitroso-β-naphthol. The structural formula is indicated below, and it appears likely that the iron atom is complexed with three naphthol radicals, only one of which is shown:

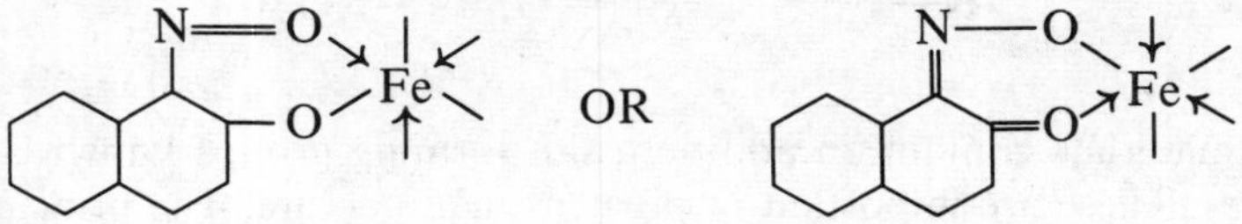

In the pure state, Pigment Green B is an intense dark but dull green. When reduced with white, or blended, it produces pleasant green shades. The alkali resistance is good and it is therefore used in distempers and water paints.

Nitro group

This group is now almost obsolete but is included because the method of preparation of the best-known member, Lithol Fast Yellow GG, is of interest.

Lithol Fast Yellow GG is made by a condensation reaction between one molecule of formaldehyde and two molecules of p-chloro-o-nitraniline, giving the compound:

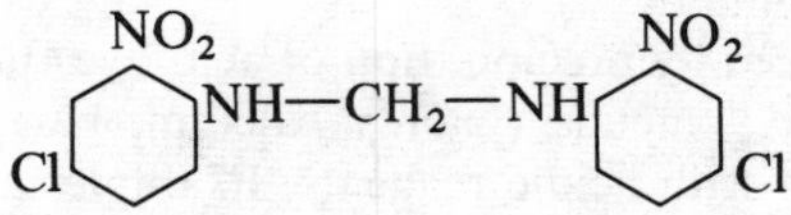

Such a reaction is not common in pigment manufacture, but condensation reactions are used in the synthesis of large molecules of high lightfastness (further details are given later in this chapter).

Lithol Fast Yellow GG was used as a stainer for water paints, but it has now been replaced by the Hansa yellows.

AZO GROUP

This is the largest group, in respect of both numbers and quantity used in the paint industry. The members all contain the chromophore $-N=N-$ and can be pigment dyestuffs, toners, or lakes.

Preparation of azo pigments

The process involves two stages (i) the conversion of an aromatic primary amine into a diazonium compound by diazotization, and (ii) the coupling of the diazonium compound with a naphthol or second amino compound to give the dyestuff or pigment.

Diazotization

When an aromatic primary amine is treated with nitrous acid in the presence of excess mineral acid at 0°C (32°F), the following reaction takes place:

$$ArNH_2 + O=N-OH + HCl \rightarrow Ar\overset{+}{N} \equiv \overset{-}{N}Cl + 2H_2O$$

Diazonium compound

The nitrous acid is produced *in situ* by addition of sodium nitrite solution to the solution or suspension of amine in mineral acid. Most diazonium compounds are unstable above 5°C (41°F) and so the temperature must be kept as near 0°C as possible. Ice is usually added for this purpose.

According to Hughes, Ingold and Ridd [1] the most probable mechanism involves the formation of nitrosyl chloride from the nitrous and hydrochloric acids:

$$O=N-OH + HCl \rightarrow O=N-Cl + H_2O$$

This then reacts with the aromatic amine to give the diazonium compound

$$ArNH_2 + O=N-Cl \rightarrow Ar\overset{+}{N} \equiv \overset{-}{N}Cl + H_2O$$

The rate of diazotization is largely controlled by the strength of the amine and this is reflected in the solubility in the acid solution. Strong bases, such as aniline and toluidine, dissolve readily and are diazotized rapidly. Weaker bases, such as 3-nitro-4-toluidine, are diazotized as suspensions in mineral acids. They dissolve slowly as the diazotization proceeds to give soluble diazonium compounds. Very weak bases, e.g. 2,4-dinitraniline, will not diazotize in dilute mineral acid and a special method has to be employed (*below*). Special methods are also used in cases where both base and diazonium compound are insoluble in dilute acid. Tobias (2-naphthylamine–1-sulphonic) acid is of this nature and the diazotization is described later.

In practice, one or other of the following methods is used.

Direct method. The base is either dissolved or suspended in dilute hydrochloric acid at 0°C (32°F), sodium nitrate solution added, and the whole stirred until the reaction is complete. This is indicated by the absence of blue coloration when spotted on starch/potassium iodide paper.

Nitrosyl sulphuric acid method. Very weak bases such as 2,4-dinitraniline are dissolved in concentrated sulphuric acid together with sodium nitrite. Diazotization takes place by the action of nitrosyl sulphuric acid:

$$NO_2\langle\bigcirc\rangle NH_2 + O=NHSO_4 \longrightarrow NO_2\langle\bigcirc\rangle \overset{+}{N} \equiv \overset{-}{N}HSO_4 + H_2O$$
NO_2   NO_2

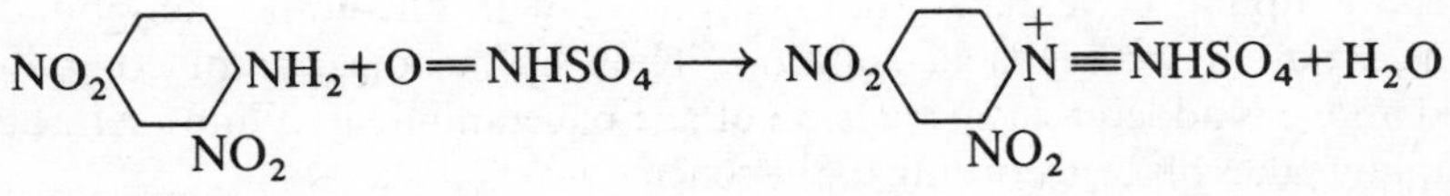

The mixture is poured into ice and water to give a soluble diazonium compound.

Reverse method. This is used when both base and diazonium compound are insoluble in water, as with Tobias acid (mentioned above). The acid is dissolved in sodium hydroxide solution and the sodium nitrite added. The mixture is then added slowly to hydrochloric acid and ice at 0°C, when the insoluble diazonium compound is formed. The insoluble diazonium salts are more stable than the soluble type and will withstand filtration and washing followed by redispersion before coupling.

Coupling

When a solution of a diazonium compound, such as that made from 3-nitro-4-toluidine (*above*) is added to a weakly alkaline solution of 2-naphthol (also called β-naphthol), coupling takes place in the 1-position by electrophilic substitution:

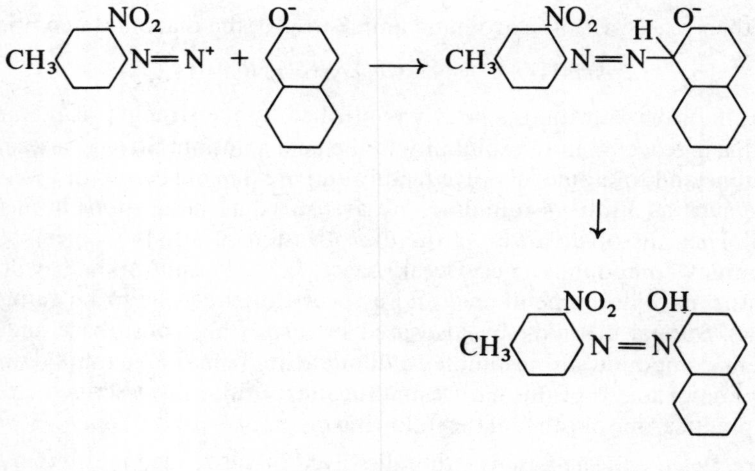

The product is the brilliant red pigment known as Helio or Toluidine Red (CI Pigment Red 3). 2-naphthol will couple slowly in slightly acid solution if it is precipitated in a fine condition immediately before coupling.

The position in which coupling takes place is determined by the recognized mechanisms. With 2-naphthol, no coupling takes place if the 1-position is occupied. With 1-naphthol, coupling takes place in the 4-position, but in the 2-position if position 4 is occupied.

The coupling of a diazonium compound with an amino compound is carried out in weakly acid solution. Thus, if the diazonium compound used above is added to a suspension of acetoacetanilide in dilute acetic acid, coupling takes place according to the scheme

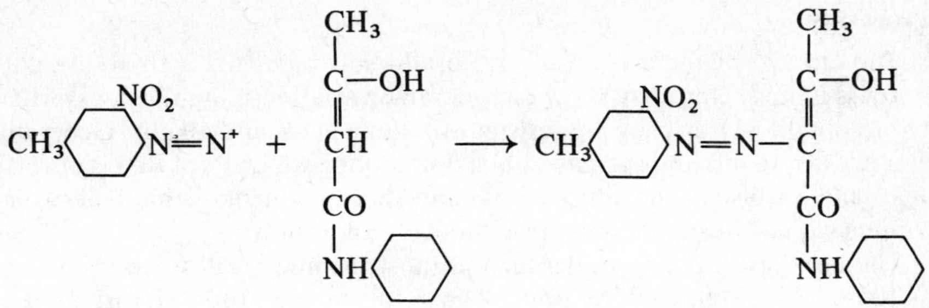

$$
\begin{array}{c}
\text{CH}_3 \\
| \\
\text{C—OH} \\
\| \\
\text{CH}_3\langle\ \rangle\text{N}{=}\text{N}^+ \ + \ \text{CH} \quad \longrightarrow \quad \text{CH}_3\langle\ \rangle\text{N}{=}\text{N—C} \\
| \\
\text{CO} \\
| \\
\text{NH}\langle\ \rangle
\end{array}
$$

The product is the bright reddish-yellow pigment dyestuff, Arylamide Yellow G, also known as Hansa Yellow G (CI Pigment Yellow 1).

RED AZO PIGMENT DYESTUFFS

The Toluidine Red described above is a typical member of this class and is one of the most widely used.

An extensive range of reds of varying shade and lightfastness is made by selection of base and coupling component. Thus Paranitraniline Red ("Para" Red) (CI Pigment Red 1) and Permanent Red 2G (CI Pigment Orange 5) are made from diazotized 4-nitraniline and 2,4-dinitraniline respectively, coupled with 2-naphthol.

A further group of azo pigment dyestuffs is obtained by coupling diazotized amines with the anilides or toluidides of β-oxynaphthoic (2-naphthol-3-carboxylic) acid (generally referred to as BONA). The latter are known as Naphthol AS or Brenthol bases and have the general formula

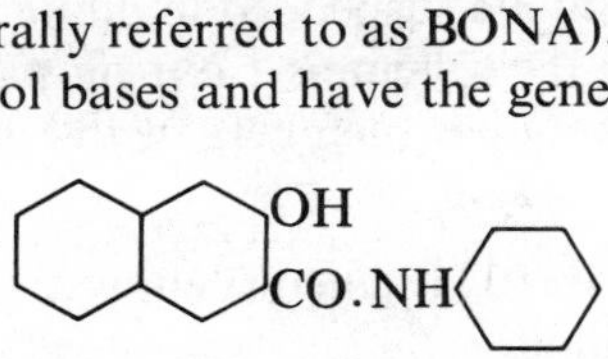

$$\text{OH} \quad \text{CO.NH}\langle\ \rangle$$

These bases couple in a similar way to β-naphthol and produce pigments of good lightfastness and which are generally deeper in shade than those made from β-naphthol. Some of the best-known members of this class are

Permanent Red F4RH (CI Pigment Red 7), made from 5-chloro-2-toluidine and BONA 5-chloro-2-toluidide.

Permanent Red FRL (CI Pigment Red 9), made from 2:5-dichloraniline and BONA p-toluidide.

Permanent Bordeaux F2R (CI Pigment Red 12), made from 5-nitro-2-toluidine and BONA-o-toluidide.

Properties

The red azo pigment dyestuffs are bright soft pigments with good light-fastness in mass, but they show considerable variation in lightfastness when reduced with white. They are resistant to dilute acids and alkalis. Generally they are fast to oils and solvents, but a few, among which Para Red is notorious, tend to bleed, and films containing these pigments cannot be over-painted with white or pale colours without discoloration.

The heat resistance is moderate and most members will tolerate stoving at 120°C (248°F) for half an hour. Above this temperature, performance is irregular; for example, Toluidine Red will tend to volatilize or turn brown.

These red azo pigment dyestuffs are hydrophobic and flush readily from aqueous dispersion into organic media. If, therefore, they are to be used in emulsion paints they should be converted to the hydrophilic condition by treatment with an anionic or suitable non-ionic surface-active agent.

The specific gravities of these pigments are in the range 1·4 to 1·5.

YELLOW AND ORANGE AZO PIGMENT DYESTUFFS

This is a smaller group than the red, but the members are bright colours with good tinting strength. They are non-toxic and in recent years the yellow members have replaced lead chromes in many types of paint. The following are the principal members of the group.

Arylamide (Hansa) yellows

These are prepared by coupling diazotized amines with acetoacetanilide or its derivatives under weakly acid conditions. Thus Arylamide Yellow G (CI Pigment Yellow 1) is prepared from diazotized 3-nitro-4-toluidine by coupling with acetoacetanilide suspended in dilute acetic acid. The reaction takes place according to the scheme set out on page 129. It is the reddest of the arylamide yellows. Other members of the group of interest in paint manufacture are

> Arylamide Yellow 5G (CI Pigment Yellow 5) from 2-nitraniline
> and acetoacetanilide
> Arylamide Yellow 10G (CI Pigment Yellow 3) from 4-chloro-
> 2-nitraniline and acetoacet-2-chloranilide.

Properties of the arylamide yellows

Generally there is a gradation in properties from the reddest and strongest Yellow G to the greenest and weakest Yellow 10G. The lightfastness also increases in passing from the G to the 10G. This change from redder to greener tone accompanied by an increase in lightfastness is associated with the introduction of chlorine into the pigment molecule. This effect of halogen introduction is fairly general in organic pigments.

The arylamide yellows are all bright and soft pigments with good fastness to light. The resistance to heat is moderate, but they withstand conventional stoving schedules, e.g. $\frac{1}{2}$ hour at 120°C (248°F). With certain types of stoving media, some members have a tendency to form a crystalline "bloom" on the stoved surface. This can be brushed off.

They are hydrophobic and flush readily from aqueous dispersion into organic media. When used as stainers for emulsion paints they must therefore be converted to the hydrophilic condition by treatment with an anionic or suitable non-ionic surface-active agent. They are also used as stainers in solvent-based paints and for making very bright greens with phthalocyanine blue.

The specific gravities of the arylamide yellows are in the range 1·4 to 1·5.

Benzidine yellows

These are prepared by coupling tetrazotized benzidines with aceto-acetanilide or its derivatives. The most commonly used base is 3:3′-dichloro-benzidine. This is tetrazotized and coupled with acetoacetanilide according to the scheme

Benzidine Yellow (CI Pigment Yellow 12)

Properties

These yellows possess fair opacity and appreciably greater tinting strength than the Arylamide yellows. The lightfastness varies with the nature of substituent atoms, the chlorine substituted types being the best.

Pyrazolone type. These are made from benzidines and derivatives of Pyrazolone. The best-known member of the class is Permanent Orange G.

Permanent Orange G is made by coupling one molecule of tetrazotized 3:3′-dichlorobenzidine with two molecules of 1-phenyl-3-methyl-pyrazolone:

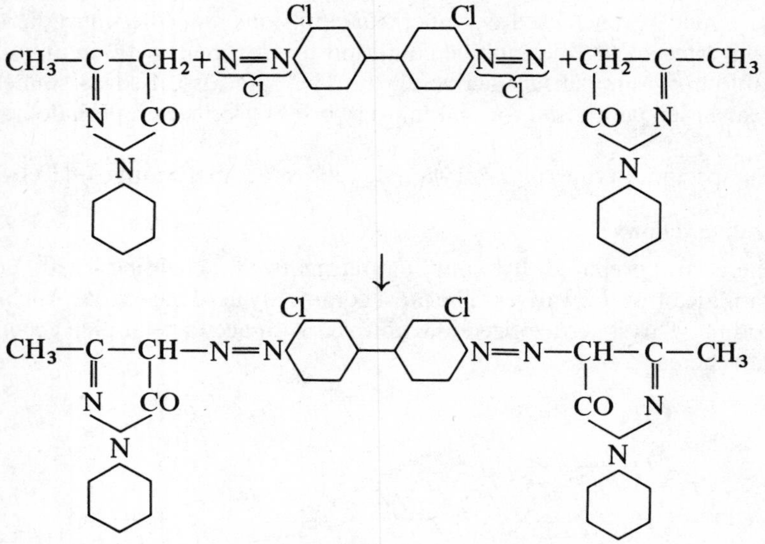

Permanent Orange G (CI Pigment Orange 13)

Properties. A very bright orange pigment dyestuff with good lightfastness but poor opacity.

AZO CONDENSATION PIGMENTS

In general the lightfastness of azo pigments, as well as their resistance to heat and solvents, tends to improve with increase in size and complexity of the molecule. The synthesis of these large molecules proved difficult. For example, disazo compounds were not readily produced on attempting to react a coupled product with further diazonium salt [2]. Success was achieved by Geigy who devised a method for combining two mono-azo pigment molecules by condensation. Azo derivatives of 2-oxynaphthoic acid can be converted to acid chlorides, and these can then be condensed with diamines to produce a large molecule. The scheme on facing page illustrates the general principle.

A range of azo pigment dyestuffs of this and other types are now being produced. They show, in general, good lightfastness and resistance to heat and a reduced tendency to bleed in solvents.

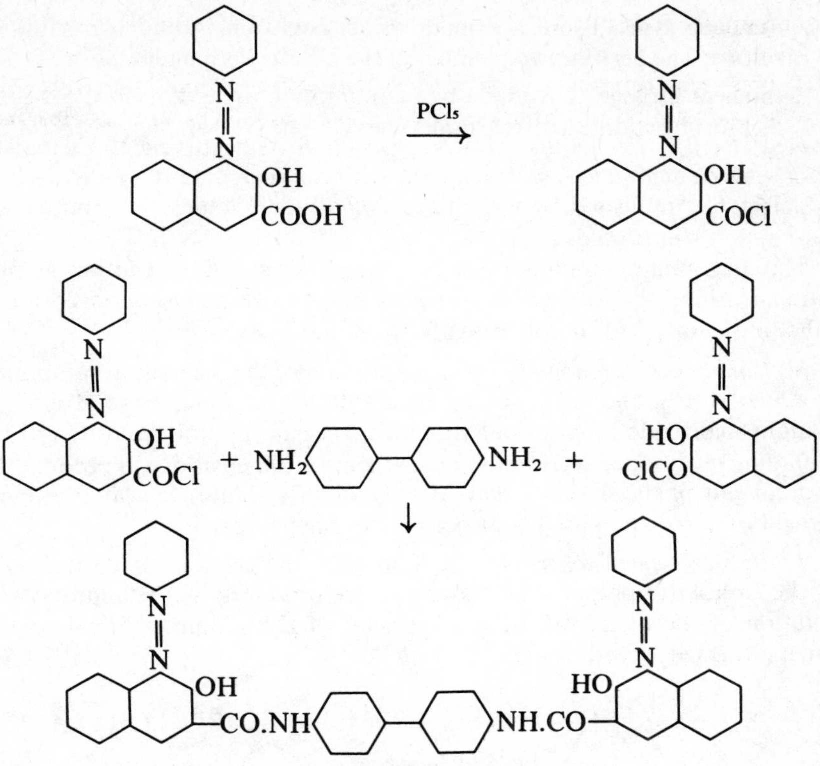

Azo condensation pigments

AZO TONERS AND LAKES

These all contain one or more salt-forming groups of the acidic type such as sulphonic ($-SO_3H$) and/or carboxyl ($-COOH$). These groups can be present in either or both of the coupling components. The best known member of the Toner class is **Lithol Red R** (CI Pigment Red 49) which is made by coupling diazotized β-naphthylamine-α-sulphonic (Tobias) acid with β-naphthol. The reaction follows the scheme

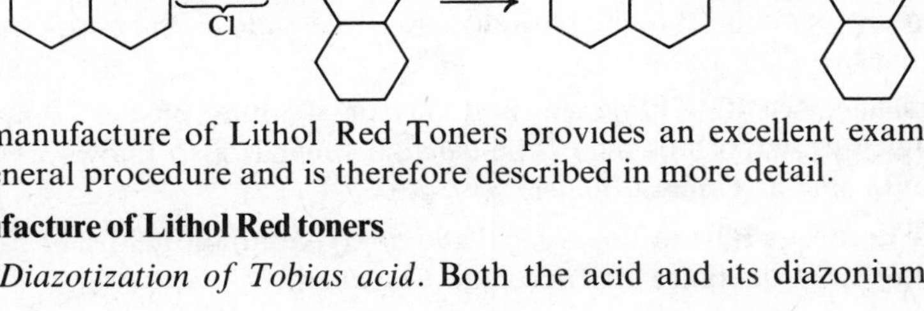

The manufacture of Lithol Red Toners provides an excellent example of the general procedure and is therefore described in more detail.

Manufacture of Lithol Red toners

(i) *Diazotization of Tobias acid.* Both the acid and its diazonium compound are insoluble in water and so the method of reverse diazotization is employed.

Tobias acid (1 mol) is dissolved in the necessary sodium hydroxide solution and a solution of sodium nitrite (1 mol) added. The mixed solution is cooled to 10°C (50°F) and added slowly and with stirring to $2\frac{1}{2}$ mol of hydrochloric acid in ice and water. The whole is stirred until no nitrous acid is detectable, indicated by the absence of a blue colour when spotted on starch–potassium iodide paper.

The diazonium compound can be used either as the suspension as prepared or, since it is fairly stable, it can be filtered, washed to remove soluble salts, and re-dispersed in cold water for coupling.

(ii) *Coupling with β-naphthol.* The suspension of the diazonium compound is added slowly and with stirring to a solution of β-naphthol (1 mol) in sodium hydroxide (1 mol) and sodium carbonate ($\frac{1}{2}$ mol) at 10°C (50°F). Coupling takes place according to the scheme outlined above to produce the sodium salt of the dyestuff. This is only slightly soluble in water and can either be filtered off or used in suspension for further reaction.

(iii) *Conversion to toners.* Lithol toners are the calcium or barium salts of the Lithol dye prepared as above and are produced by the addition of a solution of the metal salt to a suspension of the sodium derivative. The toner is formed according to the scheme

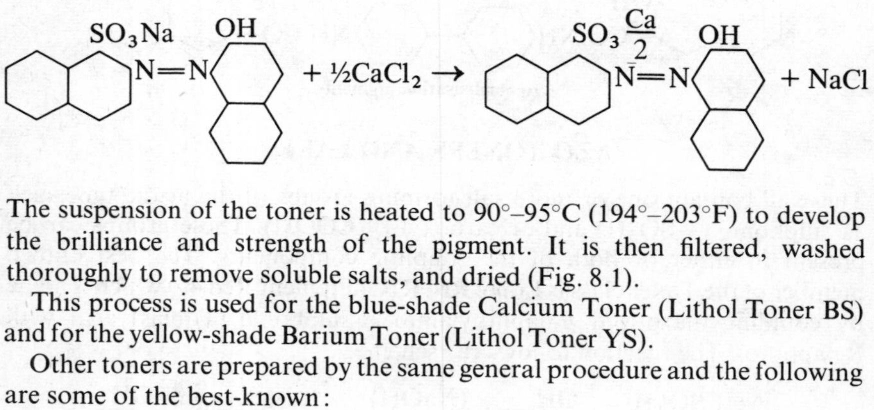

The suspension of the toner is heated to 90°–95°C (194°–203°F) to develop the brilliance and strength of the pigment. It is then filtered, washed thoroughly to remove soluble salts, and dried (Fig. 8.1).

This process is used for the blue-shade Calcium Toner (Lithol Toner BS) and for the yellow-shade Barium Toner (Lithol Toner YS).

Other toners are prepared by the same general procedure and the following are some of the best-known:

Permanent Red 2B (CI Pigment Red 48) from 2-chlor-4-aminotoluene-5-sulphonic acid and β-oxynaphthoic acid. It is made as the barium or calcium toner.

Permanent Red 4B (CI Pigment Red 57) from 4-aminotoluene-3-sulphonic acid and β-oxynaphthoic acid. The calcium toner is also known as **Lithol Rubine BK** and as **Crimson Toner.**

Lake Bordeaux B from Tobias acid and β-oxynaphthoic acid. The calcium and manganese toners are used in cellulose lacquers.

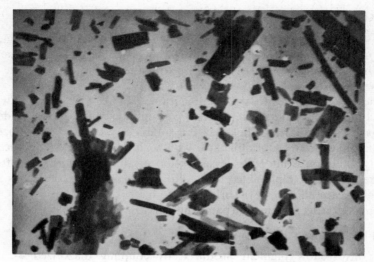

Courtesy: Paint Research Association

Fig. 8.1   Photomicrograph of Lithol toner (× 12 000)

General properties of Azo toners

The opacity of these pigments is superior to that of the pigment dyestuffs and they possess high tinctorial strength. With the exception of the Lithol toners, which are poor, the lightfastness of the class is fairly good but is not up to that of the pigment dyestuffs. In pale tints they all tend to fade. Generally they show good heat and solvent resistance and are used in air-drying and stoving paints.

Rosinated toners

If an alkaline solution of rosin (virtually a solution of sodium abietate) is added to the suspension of sodium lithol before addition of the heavy metal salt, the precipitated pigment will consist of an intimate mixture of the Lithol toner and an abietate or rosinate of the heavy metal. Such materials are known as Rosinated toners and can contain from 10 to 30 percent of metal rosinate. They show a greater degree of brilliance than the straight toner and are of comparable strength.

Rosinated Lithol toners are used in printing inks for the following reasons: (i) brilliance of self-colour, (ii) softness of texture, (iii) ease of dispersion in oils and other organic media, and (iv) moderate cost. They are seldom used in paints since Lithols lack the necessary fastness to light.

A number of other toners are also made in rosinated forms, e.g. **Permanent Red 2B** and **Permanent Red 4B.**

Lakes from azo colours

Many of these are reduced toners and can be prepared by forming the toner in presence of the base. The bases used include barytes, blanc fixe, alumina, alumina/blanc fixe, china clay and alumina/china clay. The ratio of base to toner can be varied at will.

Other lakes are formed by combination of a reactive dyestuff with a base containing alumina. An interesting example of this type is afforded by **Pigment Scarlet 3B** where the dyestuff is soluble in water and is produced by coupling diazotized anthranilic acid with β-naphthol-3:6-disulphonic acid. The dye solution is added to a suspension of alumina/blanc fixe, often in the presence of a sulphonated oil and a zinc soap. The process has to be controlled carefully to give the brightest product.

Pigment Scarlet 3B is used in paint and cellulose lacquers.

Nickel Azo Yellow (CI Pigment Green 10)

An olive-yellow pigment produced by coupling diazotized 4-chloro-aniline with 2,4-dihydroxyquinoline followed by conversion to the nickel co-ordination compound. It has the following constitution:

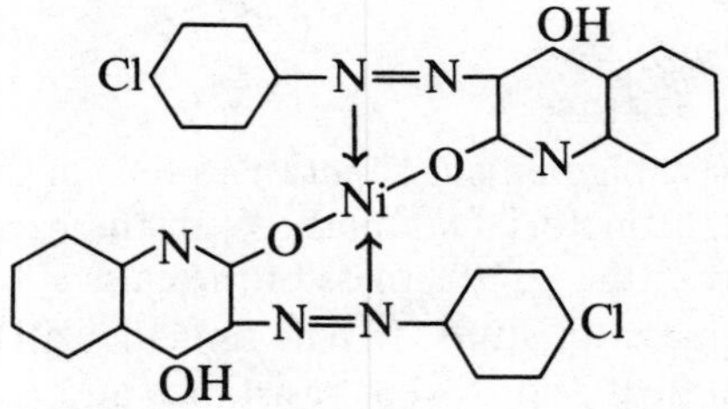

It possesses good fastness to light and heat and to most organic solvents.

PHTHALOCYANINE PIGMENTS

The phthalocyanines comprise an important class of pigments ranging in colour from deep blue to yellowish green. The first member of the class was observed as an impurity in a batch of phthalimide being made at the works of Scottish Dyes Ltd in 1928. The material was examined by Linstead [3] who established its constitution as an iron phthalocyanine. The corresponding compounds of other metals were synthesized, and the most interesting was a deep-blue material containing copper. In the compound the atom of copper is linked to form a 16-membered ring of alternate carbon and nitrogen atoms. The copper atom is combined with the four adjacent nitrogen atoms by two covalent and two co-ordinate bonds:

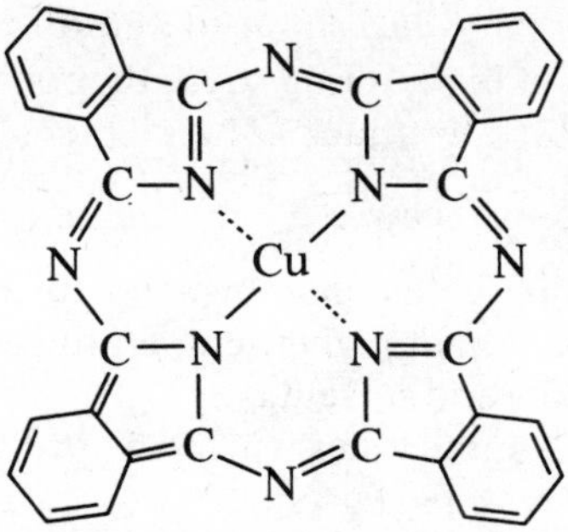

Phthalocyanine Blue (CI Pigment Blue 15)

The structure shows a molecule in which the only deviation from symmetry is the presence of one *o*-quinonoid ring. This symmetry, combined with resonance within the relatively complex molecule, contributes to the outstanding chemical and physical stability.

Many other metals have been used in place of copper, but the compounds lack the brilliancy and chemical stability of the copper compound. The latter possesses a clean brilliant tone, high tinting strength and lightfastness, and is very inert chemically. It forms the basis of the phthalocyanine range.

Preparation of copper phthalocyanine

Several methods are available, but there are two main types, (a) the dry (fusion) process and (b) the wet (solvent) process.

In the dry process, phthalonitrile and cuprous chloride are heated together in the presence of metallic copper. Phthalimide is obtained as a byproduct. Other dry processes involve heating phthalic anhydride with a copper salt in the presence of urea.

The wet process consists in heating phthalic anhydride, cupric chloride, urea and a catalyst in trichlorobenzene.

The crude copper phthalocyanine obtained by these methods is further processed to produce the pigmentary form. This process consists either in dissolving in cold concentrated sulphuric acid followed by pouring into hot water to precipitate the blue, or in milling with sodium chloride in presence of a solvent. Both processes yield a product of fine particle size and different crystal form from the crude.

Properties of copper phthalocyanine

Copper phthalocyanine exists in several crystal forms, but two only are of practical interest. These are designated as α (the redder) and β (the greener) forms. The crude compound possesses the β-form but is converted to the α-modification on further processing [4]. Early α-forms tended to revert to the larger crystals when in contact with certain aromatic solvents, but this fault has now been overcome and stable α- and β-forms are on the market.

It is a bright pigment of high tinting strength, poor opacity, and outstanding stability and resistance to attack. The pigment is resistant to all acids, with the possible exception of concentrated sulphuric acid in which it is soluble but is reprecipitated on dilution.

Copper phthalocyanine withstands heat up to 200°C (392°F) and sublimes without decomposition at higher temperatures. It is fast to light and is insoluble in all paint media and solvents.

A fault with phthalocyanine blues which has persisted over many years is the tendency to flocculate in paint systems. Flocculation is the formation of clusters of pigment particles after dispersion and can occur in the paint can and in the wet paint film. This fault can be very troublesome if the blue is used to tint a white enamel. Application by brush can break down the flocculates, resulting in blue streaks, while spraying can result in a mottled finish. The introduction of non-flocculating blues has largely corrected this fault and also the tendency to flotation.

The following are the physical characteristics of phthalocyanine blue:

> Specific gravity 1·6   pH of aqueous extract 7·0
> Oil absorption 46

Uses of phthalocyanine blues

Their outstanding properties enable them to be used in paints, inks, artists' colours and, in fact, wherever stable blues are required. For this reason they have largely displaced Prussian blue. Very bright greens are produced when they are mixed with arylamide yellows.

Phthalocyanine greens (CI Pigment Green 7 and 36)

The four aromatic rings in the copper phthalocyanine molecule contain 16 replaceable hydrogen atoms. These can be replaced progressively by chlorine or a mixture of chlorine and bromine to give a range of greens. Chlorine alone gives bluish greens, but yellowish greens result when both chlorine and bromine are used to replace hydrogen atoms; for example, 12 hydrogen atoms can be replaced by 8 of chlorine and 4 of bromine or 15 hydrogen atoms by 3 of chlorine and 12 of bromine.

The greens are similar to the phthalocyanine blues in their outstanding resistance properties and are used for similar purposes.

VAT PIGMENTS

Vat dyestuffs are among the fastest to light of textile dyes and some of these have been found to be very useful pigments. The best-known member of the class, **Indigo**, has, however, been little used as a pigment as it lacks brilliance. The interesting members of the class include the following:

THIOINDIGO DERIVATIVES

These comprise a range of colours from pink through red to violet and they all contain chlorine atoms in the molecule. Thus, the constitution of Thioindigo Bordeaux is as follows [4]:

Thioindigo Bordeaux (CI Pigment Red 88)

This pigment is very fast to light and solvents, but within the group there is appreciable variation in these properties.

ANTHRAQUINONE PIGMENTS

Alizarin (1:2-dihydroxyanthraquinone) is a material which has been known for a long time. It was originally prepared from madder root (*Rubia tinctorum*) in which it occurs as a glucoside, but it is now prepared from the parent anthraquinone. Alizarin yields intensely coloured compounds with a number of metals, but the straight toners are of little interest as they possess poor working properties. More useful pigments are obtained by formation of lakes and these are precipitated on to alumina, usually in the presence of a surface-active agent such as Turkey Red Oil (sulphonated castor oil).

Synthetic Madder Lake (CI Pigment Red 83) is the calcium salt of alizarin precipitated in this way. It is very fast to light and until comparatively recently was used as a standard for lightfastness. Today it is used in artists' colours but finds little use in the paint industry.

Other anthraquinone pigments

A wide range of colours is obtained from the more complex anthraquinone derivatives. Some of these are very fast to light and the constitutions of two of the best-known, **Flavanthrone Yellow** and **Indanthrone Blue**, are depicted overleaf.

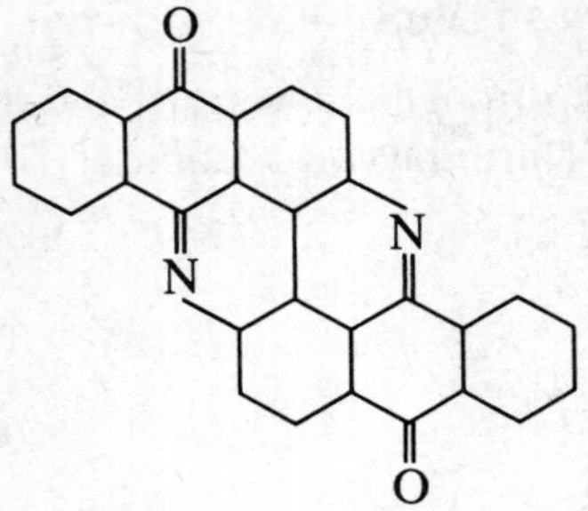

Flavanthrone Yellow
(CI Vat Yellow 1)

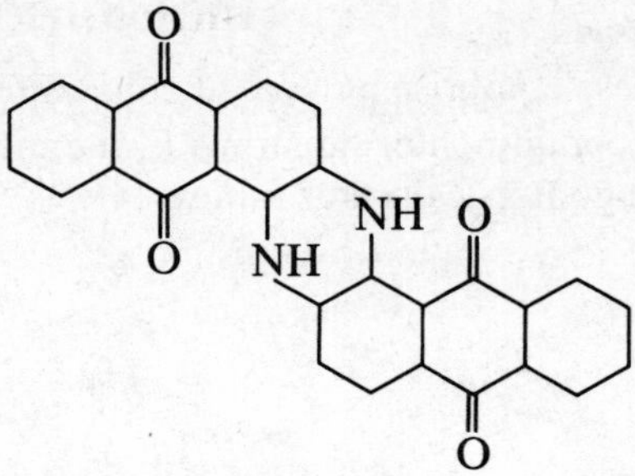

Indanthrone Blue
(CI Vat Blue 4)

Although the lightfastness of some of these pigments approaches that of phthalocyanine blue, some members exhibit a tendency to bleed slightly in strong solvents.

Still greater complexity of molecular structure occurs in the Perinone and Perylene pigments. Perinone Orange (*below*) possesses very good fastness to light, solvents and heat and is used in car finishes.

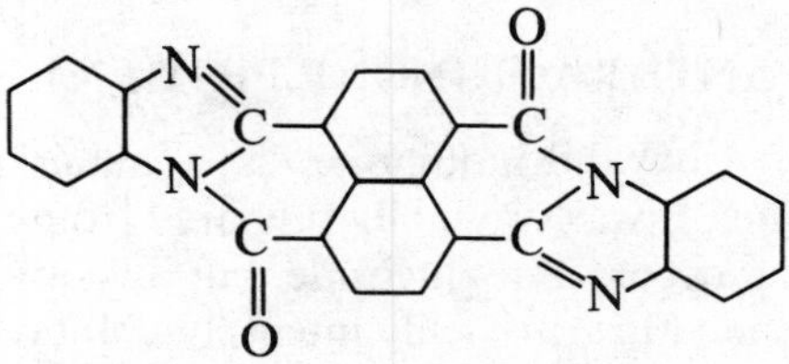

Perinone Orange
(CI Vat Orange 7)

The perylenes are also fast to light and some approach the lightfastness of phthalocyanine blue. They also have excellent heat resistance and for this reason are used in plastics. The constitution of a typical member of the class, Perylene Maroon, is depicted below:

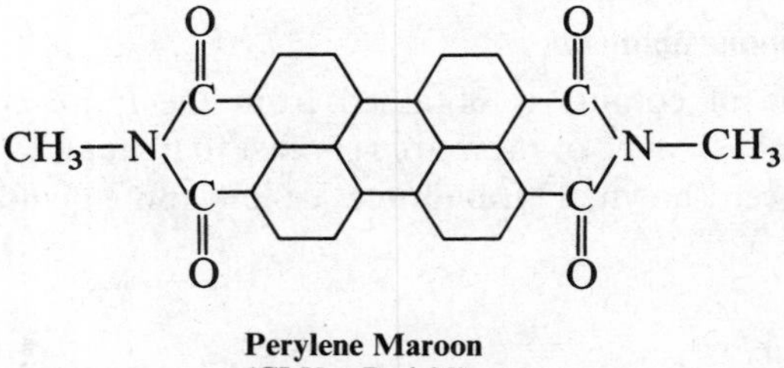

Perylene Maroon
(CI Vat Red 23)

QUINACRIDONE PIGMENTS (CI Pigment Violet 19)

The polycyclic quinacridone molecule (*below*) exists in three stable crystalline forms, two red (*a* and *γ*) and one reddish-violet (*β*). These form the basis of the range of pigments.

The compound was synthesized by du Pont [5] by condensing a dialkyl succinylsuccinate with aniline, followed by ring closure and oxidation:

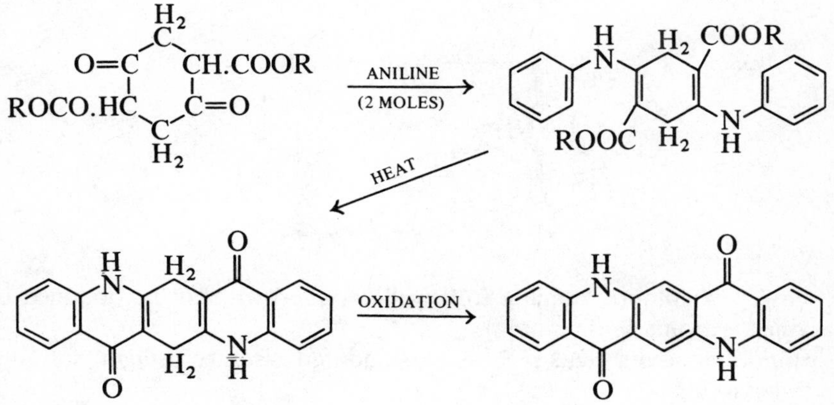

$$ \text{ANILINE (2 MOLES)} \longrightarrow \qquad \xrightarrow{\text{HEAT}} \qquad \xrightarrow{\text{OXIDATION}} $$

Modifications to the basic shades are made by introducing substituents.

The quinacridone pigments possess a lightfastness comparable with that of the phthalocyanine, together with high tinting strength. They are therefore valuable where clean fast tints are required, as on cars and coach work generally.

DIOXAZINE PIGMENTS

The pigments of this class are clean violet shades, and one member, carbazole dioxazine violet (CI Pigment Violet 23), is well known. It has the following constitution:

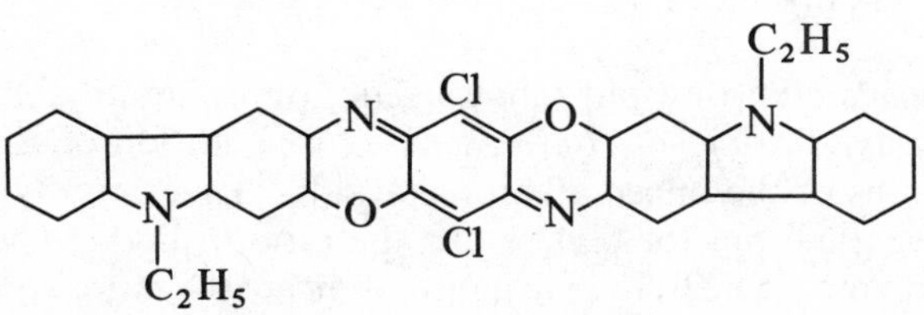

This pigment possesses very high tinting strength and good lightfastness both in self colour and in tints. It has a very red tone and is useful for shading the greenish phthalocyanines.

ISOINDOLINONE PIGMENTS

These form a range of compounds produced by condensing two molecules of tetrachloroisoindolinone with an aromatic diamine. They possess the following basic structure:

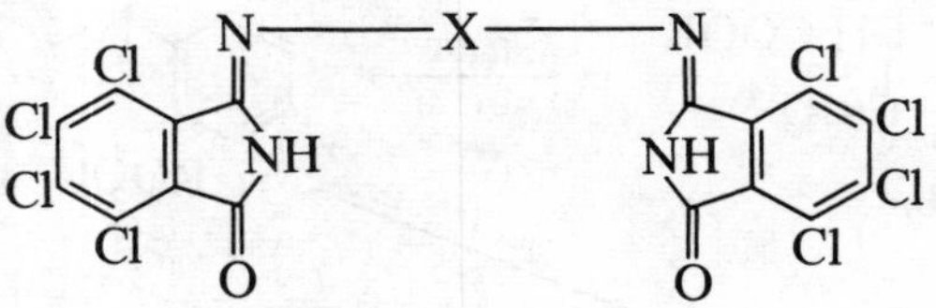

A variety of colours, ranging from yellow to brown, can be obtained by varying the aromatic diamine [6].

Isoindolinone pigments possess outstanding resistance to light, solvents, and weathering.

PIGMENTS FROM BASIC DYES

Historical. The conversion of basic dyestuffs into pigments by precipitation from acid solutions has been practised for many years. Precipitants used have included sodium phosphate and tannic acid, either alone or with tartar emetic. Lakes were generally made on an alumina base and were very brilliant but possessed poor lightfastness.

With the introduction of the so-called complex acids it was found possible to produce brilliant colours with good tinting strength and much better lightfastness. These were first produced in Germany under the name "Fanal" colours.

FANAL TYPE PIGMENTS

These are toners produced by precipitation of basic dyestuffs with complex acids.

Complex acids comprise phosphotungstic, phosphomolybdic and phosphotungstomolybdic acids and are prepared by addition of strong mineral acid to solutions of the mixed alkali metal salts. Theoretically any proportions could be used, but for best results the ratio of P_2O_5 to MoO_3 is kept within certain limits. The following figures indicate the ratios employed:

Phosphotungstic acid	Commonest ratio $P_2O_5/24WO_3$
	Others used are 1/20, 1/16 and 1/12
Phosphomolybdic acid	Commonest ratio $P_2O_5/24MoO_3$
	Others used are 1/22, 1/20 and 1/18
Phosphotungstomolybdic acid	Ratios used are $P_2O_5/10WO_3/8MoO_3$
	and $P_2O_5/9WO_3/9MoO_3$

For best results there would appear to be a relationship between the number of free acid groups in the complex acid and the number of basic groups in the dyestuff.

The following basic dyestuffs are commonly used:

Methyl Violet	Auramine
Victoria Blue	Rhodamines
Brilliant Green	Setoflavine

In the manufacture of the toner, a salt of the basic dyestuff is dissolved in water and a solution of the complex acid added. The brightness and tinctorial strength of the pigment depend on accurate control of pH, temperature and concentration.

Phosphotungstomolybdic acid is the most commonly used acid and the products are therefore described as *triple salt toners*.

On account of their brilliance and high tinctorial strength the Fanal type colours are used widely in printing inks, but the lightfastness is not quite good enough for use in paints.

PIGMENTS FROM ACID DYESTUFFS

The acid dyestuffs employed are mostly of the azo and xanthene type and are very soluble in water. They are often difficult to precipitate as toners, and generally an absorbing base such as alumina aids complete precipitation. The acid dyestuffs in common use include Pigment Scarlet 3B, eosin, fluorescein, acid greens and erioglaucine. They are precipitated as salts of the heavy metals calcium, barium or lead.

Properties

The lakes produced from acid dyestuffs are very brilliant and possess high tinting strength. However, the lightfastness is poor and they are very sensitive to alkali. With the exception of Reflex Blue (*below*) and Pigment Scarlet 3B they are now little used in the coating industry.

Reflex Blue

This is prepared from Alkali Blue which is the sodium salt of the monosulphonic acid of phenylated rosaniline. The latter is dissolved in water and the Reflex Blue precipitated under controlled conditions with dilute acid. The product is filtered, washed, and flushed into lithographic varnish.

Reflex Blue possesses high tinting strength and moderate lightfastness. It is used for toning black inks and in metal printing where its good heat-resistance is an asset.

BASES FOR LAKES

Alumina, $Al_2(OH)_6$, is the principal base used for the best-quality lakes. It is prepared by addition of sodium carbonate solution to a solution of aluminium sulphate:

$$Al_2(SO_4)_3 + 3Na_2CO_3 + 3H_2O \rightarrow Al_2(OH)_6 + 3CO_2 + 3Na_2SO_4$$

Alumina is amphoteric and possesses a very soft texture. It is fairly transparent.

Blanc Fixe is a non-reactive base sometimes used for good-quality lakes. The lake is precipitated on to a suspension of the base. Blanc Fixe is sometimes used in conjunction with alumina under the name Alumina/Blanc Fixe.

China Clay forms a permanent suspension in water and is often used as a base for lakes to be supplied in paste form. It is sometimes used with alumina under the name Alumina/China Clay.

Toxicity of organic pigments. The majority are accepted as being reasonably safe. Possible exceptions are barium toners which may contain soluble barium. In this case the necessary measures must be taken to protect the operatives, and a high standard of industrial hygiene must be maintained.

References to Chapter 8

[1] HUGHES, E. D., INGOLD, C. K., & RIDD, *Nature*, 1960, **166,** 642.
[2] GAERTNER, H., Modern chemistry of organic pigments. *JOCCA*, 1963, **46,** 13.
[3] LINSTEAD, R. P., *J. Chem. Soc.*, 1934, p. 1016.
[4] SMITH, F. M. & EASTON, F. D., Phthalocyanine pigments—their form and performance. *JOCCA*, 1966, **49,** 614.
[5] ALLEN, R. L. M., *Colour Chemistry*. Nelson, London, 1971, p. 257.
[6] Geigy, U.K. Pat. 833,548; 1,093,669.

9 Solvents and plasticizers

Solvents are used in paints and lacquers to reduce the viscosity or consistency of the material and so facilitate the application of a uniform coating. They must be solvents that are suitable for the oil or resin present. After application the solvent is no longer required and should evaporate completely from the film.

The resins used in lacquers frequently form hard and brittle films and require plasticizers to confer the necessary elasticity and adhesion. These are usually liquids which must have a minimum tendency to volatilize so that they will remain in the film. It is essential that the plasticizer should be completely miscible with the resin, and to this extent it can be regarded as a solvent.

SOLVENTS

Paint and lacquer solvents, with the exception of water, are volatile organic liquids which can be classified into well-defined chemical groups. The solvent power and general physical properties vary over a wide range. A useful but arbitrary classification of solvents is by boiling point (though these are not necessarily related to evaporation rates), thus:

Solvents	*Boiling point*
Low boiling point	Below 100°C (212°F)
Medium boiling point	Between 100° and 150°C (212° and 302°F)
High boiling point	Between 150° and 250°C (302° and 482°F)
Plasticizers	Above 250°C (482°F)

The important properties of paint solvents are the following:

(i) Solvent power
(ii) Rate of evaporation
(iii) Boiling point and distillation range
(iv) Flash point and inflammability
(v) Toxicity.

The following properties are also of importance as indicators of purity—specific gravity, colour, moisture content, acidity or alkalinity, and non-volatile matter.

GENERAL PROPERTIES OF SOLVENTS

Solvent power

The film-forming materials used in paints and lacquers are, with a few exceptions, organic in nature and possess no crystalline structure. When treated with suitable solvents they dissolve to give solutions the viscosity of which increases with increasing concentration. There is no solubility limit, as in inorganic solutions, but problems of application limit the concentrations used. These solutions produce continuous films on evaporation of the solvent.

It is well known that certain types of resin or polymer dissolve only in certain types of solvent ("like dissolves like") and it was early realized that the polarity of the molecules of each was an important factor. Polar compounds are unsymmetrical in molecular structure and can contain groups which cause changes in the electron distribution. Typical examples are the nitro ($-NO_2$), carbonyl ($=CO$) and hydroxyl ($-OH$) groups. Molecules containing such groups exhibit a permanent dipole moment, indicating the presence of separated charges.

One result of this polarity is the tendency of such molecules to form associated groups by attraction of opposite poles. A consequence of this association is the fact that the viscosity of polar liquids is generally greater than that of non-polar liquids of similar molecular weight.

The effect of polarity in the solute is illustrated by cellulose nitrate which dissolves readily in polar solvents such as esters and ketones but is insoluble in non-polar hydrocarbons. On the other hand, some of the weakly polar drying oils are insoluble in polar alcohols but soluble in non-polar hydrocarbons.

In both polar and non-polar solvents there are variations in solvent power or strength between individual types. Thus in the non-polar solvents, the aromatics give solutions of resins of lower viscosity (at the same concentration) than those in aliphatic hydrocarbons.

Miscibility of solvents

When two liquids, A and B, are brought into contact, mixing will occur only if the molecules of species A are able to move freely among those of species B. If, however, the attractive forces between A and A or B and B are greater than those of A for B, then the two liquids will be immiscible.

The attractive forces between the molecules in a liquid are responsible for the internal cohesion of the liquid, and the intensity of this is known as the "cohesive energy density" [1]. It can be defined as the energy required to separate the molecules in unit volume of liquid or the cohesive energy per unit volume. It is related to the latent heat of vaporization in the following way:

$$\text{Cohesive energy density} = \frac{\Delta H - RT}{M/D}$$

where $\Delta H =$ latent heat of vaporization
  $R =$ gas constant
  $T =$ absolute temperature
  $M =$ molecular weight
  $D =$ density.

It can also be defined as the energy required to separate the molecules in 1 cm^3 of liquid, i.e.

$$\text{Cohesive energy density} = \frac{\Delta H_v}{V}$$

where $\Delta H_v =$ molar heat of vaporization in calories
  $V =$ molar volume in cm^3.

Solubility parameter

This term was suggested by Hildebrand and Scott [2] for the square root of

the cohesive energy density, i.e. $\left(\frac{\Delta H_v}{V}\right)^{\frac{1}{2}}$, and has been assigned the symbol δ.

Solvents which possess similar solubility parameters possess similar internal cohesions and, in the majority of cases, solvents are miscible, provided the solubility parameters do not differ by more than one unit. Similarly, polymers dissolve in solvents of similar solubility parameters, but, as in the case of solvents, exceptions to this general rule are encountered.

This problem was investigated by Burrell [3] who suggested that hydrogen bonding played an important part in cohesive energy density, and also that solubility parameter values should be a range rather than single figures. However, difficulties were encountered initially in assigning quantitative values to the hydrogen bonding factor and these are explained by Burrell [4] in a paper which also summarizes the applications of the solubility parameter concept. Other investigators [5, 6] have suggested that three types of intermolecular forces — dispersion, H-bonding and dipole moment — should be included in solubility figures and have proposed a "three-dimensional" solubility parameter.

Solubility parameter figures for a wide range of solvents, calculated from vapour pressure data, have been published by Hoy [7].

Solubility of polymers

Whilst the values for the solubility parameters of solvents can be calculated from heats of vaporization, polymers cannot be treated in this way. Figures for amorphous polymers are determined experimentally by ascertaining the solvent or solvent mixture in which the polymer exhibits maximum solubility. The figure assigned to the polymer is that of the solvent or solvent mixture used. Values for solvent mixtures are calculated on a volume basis. Graphical analysis of resin solubilities has been employed by Teas [8].

"Crystalline" polymers, e.g. PTFE (polytetrafluoroethylene) or polyethylene, are not amenable to the treatment described above since they are insoluble in normal solvents at ambient temperature. If they are heated above their melting points in the presence of certain solvents they will dissolve, but separate on cooling.

Summarizing, the solvent power of a solvent involves three factors— solubility parameter (δ), a hydrogen bonding factor, and dipole moment (polarity). A figure for hydrogen bonding entitled "Net Hydrogen Bonding Index" has been advanced by Nelson et al. [9]. Hydrogen bonding is low in aliphatic and aromatic hydrocarbons, moderate in ethers, esters and ketones, and high in alcohols, amines and acids.

Measurement of solvent power (traditional method)

Kauri–butanol value

This test is applied to solvents to be used in varnishes and paints. The solvent to be tested is put into a burette and added slowly to 20 g of a solution of Kauri gum in butanol (100 g Kauri to 500 g butanol). The solution is contained in a flat-bottom flask which is placed on a printed sheet of paper. The end-point is reached when the precipitated gum causes the outline of the print to appear blurred. The volume of solvent added (in millilitres) is the Kauri–butanol (KB) value. A good solvent is indicated by a high KB value.

Dilution ratio

This is important with solvents for cellulose nitrate which are divided into three classes:

(i) True solvents;

(ii) Latent solvents which are not solvents for cellulose nitrate but which increase the solvent power of the main solvent;

(iii) Diluents which are not solvents for cellulose nitrate but which are added to reduce cost or to assist the solution of other ingredients.

The dilution ratio can be defined as the tolerance of a solution of cellulose nitrate in a given solvent for a given diluent and is determined in the following way.

A solution of cellulose nitrate is made up in the given solvent and titrated with the diluent in question until a permanent slight precipitate is formed. The titration can be repeated after addition of a further quantity of solvent, but the concentration of cellulose nitrate at the end should not be less than 8 percent. The dilution ratio is the volume of diluent added divided by the volume of solvent used.

The concept of latent solvent applies also in systems other than cellulose nitrate lacquers. A mixture of xylol and n-butanol is a much more effective solvent than xylol alone for stoving finishes, especially for the alkyd–amino type.

Rate of evaporation

This is of fundamental importance in the choice of a solvent for a particular use, and with solvents of the same type, e.g. the petroleum distillates, it is roughly proportional to the distillation range. This does not hold, however, with solvents of different classes. The fundamental factors which control evaporation are the specific heat of the material and the latent heat of evaporation, and these can differ considerably in different classes of solvents. As an example, the evaporation rate of tetrachlorethane with a boiling point of 147°C (296·6°F) is considerably higher than that of water with a boiling point of 100°C (212°F).

A simple and rapid method for evaluation of evaporation rate is to place 0·5 ml of the solvent on a No. 1 Whatman filter paper. This is suspended from the edge, away from direct draughts, and the time noted for complete evaporation. Ethyl ether is frequently taken as a standard for fast evaporating solvents, but other standards, e.g. 2° xylol, can be used.

This test is known as the filter-paper spot method and is useful also for detecting the presence of oil or other non-volatile (or less volatile) contaminant. Oils and greases form a permanent spot on the paper.

It is often desirable to measure the rate of loss of solvent under conditions similar to those in a drying paint film. For this purpose a suitable instrument is the Shell Liquid Film Evaporometer [10] in which the rate of solvent loss from resins or plasticizers can be measured. The instrument also enables an analysis to be made of the solvent mixture present at various stages during the evaporation.

Distillation range

The boiling points of pure liquids at a specific pressure are constant and can be used as a method of identification. If 100 ml of such a liquid is distilled in the standard apparatus [11] the whole of the sample will distil at a constant temperature.

Commercial solvents, however, are seldom pure and when distilled by the previous method, the temperature indicated by the thermometer will rise throughout the distillation. The range of temperature shown is known as the "distillation range" and is used both as a method of identification and as an indication of the purity of a given solvent. A small piece of copper foil is usually placed in the distillation flask. The degree of staining of the foil is an indication of the amount of corrosive sulphur compounds present.

In the determination of distillation range, the temperatures taken are (i) at the appearance of the first drop of distillate, (ii) at the completion of 95 ml of distillate.

Flash point and flammability

With the exception of the chlorinated hydrocarbons, organic solvents are flammable, and a fire risk is therefore associated with their use. When these liquids are heated, the concentration of vapour in the air above will increase until it will flash on application of a flame. The lowest temperature at which this occurs is known as the flash point.

The determination of flash point is carried out by one of two standard methods. The Abel (closed cup) apparatus is used for liquids which flash below 49°C (120°F) and the method of use is described in BS 3442, Part 1: 1974. The method is now embodied in The Highly Flammable Liquids and Liquefied Petroleum Gases Regulations, 1972, Schedule 1. It is also described in BS 3900, Part 8 (Danger Classification by Flash Point) and Part 9 (Determination).

A flash tester is used for the rapid determination of flash point using small quantities (2 cm³) of material and is described in BS 3900, Parts A13 and A14.

For liquids with flash points above 49°C (120°F), the Pensky–Martens apparatus is used. This is described in BS 2839:1969.

The flash points of mixtures of two solvents of differing flash point can lie between the two or can be below either of the individual flash points.

This is exhibited by mixtures of xylol and *n*-butanol. Flash points of solvents can be raised by mixing them with small quantities of the non-flammable chlorinated solvents.

If the temperature is high enough, a mixture of flammable vapour and air can ignite without application of a flame or spark. The temperature at which this takes place is known as the *auto-ignition* temperature.

Explosive limits

Mixtures of flammable vapours and air are capable of exploding on ignition if the concentration of vapour lies within certain limits. These are known as the upper and lower explosive limits (LEL) for the solvent.

Legislation relating to solvents

Regulations covering the storage, use and transport of "Petroleum Spirit" and "Petroleum Mixtures" are laid down in the Petroleum (Consolidation) Act, 1928, which was extended by the Petroleum (Mixtures) Order, 1929, SR & O 1929, No. 993. The Act defines "Petroleum Spirit" as "crude petroleum, oil made from petroleum, or from coal, shale, peat or other bituminous substances and other products of petroleum" which has a flash point (Abel closed cup) of less than 22·8°C (73°F). "Petroleum Mixtures" are mixtures of petroleum and any other materials which have a flash point (Abel closed cup) of less than 22·8°C (73°F).

The Petroleum (Flammable Liquids) Order, 1971, applies the 1928 Act to non-hydrocarbon flammable liquids.

The labelling, storage and use of highly flammable liquids is controlled by the Highly Flammable Liquids and Liquefied Petroleum Gases Regulations, 1972. In these Regulations "highly flammable liquids" are defined as those with an Abel flash point of less than 32°C (90°F) and which also support combustion under conditions specified in the Regulations.

Other Regulations of importance are The Flammable Liquids (Conveyance by Road) Regulations and The Flammable Substances (Conveyance by Road) Labelling Regulations.

Toxicity of solvents

In addition to the fire hazard, organic solvents can be harmful to the skin as they dissolve the natural oils and so expose the skin to dermatitis and other complaints. They can also be absorbed into the system through the skin. In addition, inhalation of their vapours can result in internal disorders of varying severity.

The effect on the skin is usually caused by splashes of liquid and can be minimized or prevented by the use of barrier cream and/or the wearing of protective clothing.

Inhalation of vapour is frequently difficult to avoid completely, but note should be taken of the published [12] TLV figures which represent the maximum concentration of vapour to which an operator can be exposed continuously, day after day, without ill effects. The values vary considerably from one class of solvent to another and also among individual members within a group.

The concentration of vapour in the air will be related to the volatility, so that high boiling-point solvents and plasticizers are not likely to give dangerous concentrations.

For a more detailed treatment of toxicity as well as the fire hazard associated with paint materials the reader is recommended to the section entitled "Hazards to the Paint User" by Dr H. Brunner in Hess' *Paint Film Defects*, 3rd edn, Chapman & Hall, 1979.

Other solvent properties

Specific gravity

This can be determined by several methods. For rough determinations, a hydrometer can be used. In this case it is essential to note the temperature and to use a wide vessel to avoid *wall effects*. More accurate measurements can be made with the specific gravity bottle or the Westphal balance.

Colour

With the exception of the heavy coal-tar naphthas, solvents have very little colour. If necessary, a measurement can be made using a glass cell in the Lovibond Tintometer (Chapter 2).

Moisture

An indication of the presence of moisture can be obtained by shaking the solvent with anhydrous copper sulphate, which turns blue in the presence of moisture, or by the addition of a miscible solvent which has no tolerance for water. Examples of the latter are benzene and petroleum ether. In this test the presence of moisture is shown by the separation of droplets.

Moisture can be determined quantitatively using the Karl Fischer method [13].

Acidity or alkalinity

If the solvent is miscible with water or industrial methylated spirit, a solution in one of these is titrated with standard acid or alkali using phenolphthalein as indicator. Other solvents are shaken with neutral distilled water and the aqueous extract titrated.

Non-volatile matter

The presence of non-volatile matter is often indicated in the filter-paper spot test for evaporation rate. It is determined quantitatively by placing a known quantity of solvent in an open dish and evaporating in a steam bath or in a steam-heated oven. Any non-volatile residue can then be weighed after cooling.

CLASSES OF SOLVENTS

Petroleum hydrocarbons

A considerable number of very useful and widely used solvents are obtained either by distillation of crude petroleums or by thermal cracking of petroleum gases. The products obtained by distillation are mainly paraffin hydrocarbons containing small amounts of aromatics. The thermal cracking processes are designed to give solvents with high aromatic contents.

Low-boiling petroleum fractions

Petroleum ethers (ligroins)

These are marketed in several grades according to boiling range. The three most common ranges are 40° to 60°C (104° to 140°F), 60° to 80°C (140° to 176°F), and 80° to 120°C (176° to 248°F). The specific gravities range from 0·645 to 0·676 and the aromatic contents from 1 to 5 percent.

Special boiling-point solvents

These are produced in six or more boiling ranges between about 30° and 160°C (86° and 320°F). They are generally referred to as SBP 1, 2, 3, etc. and can vary considerably in aromatic content. The solvent power generally increases with aromatic content, and members of this range are used as rubber solvents as well as for lacquer formulation.

Flash points below 23°C (73°F): Highly Flammable Liquids.

White spirit

This solvent was introduced as a paint thinner in about 1885 by Samuel Banner and for many years it was regarded as a substitute for turpentine. The term "turpentine substitute" persists in some quarters, but the term "white spirit" is universally accepted in the British paint industry.

White spirit is produced from the petroleum fraction distilling between the low boiling fractions (the SBPs and petrols) and the burning oils. The boiling range is mainly between 150° and 190°C (302° and 374°F), although a small percentage may boil up to 210°C (410°F). This high-boiling fraction evaporates slowly at room temperature and should therefore be as small as possible. The actual composition and especially the aromatic content varies with the process and source of petroleum, but supplies generally conform to the following requirements of BS 245:1976 [11].

Boiling range	Up to 155°C (311°F)	10% max.
	Below 195°C (383°F)	90% min.
	Max. boiling point	210°C (410°F)
Flash point 34°C (93°F) minimum.		

In addition the material must be neutral, free from objectionable sulphur compounds, and the residue on evaporation must not exceed 0·005 percent.

The characteristic odour of white spirit is due mainly to impurities and aromatics. A special grade of odourless white spirit with a higher flash point (46°C, 115°F) is available, but due to the removal of aromatics the solvent power is not as good as the normal grade.

White spirit possesses moderate solvent power and is miscible with most organic solvents. It dissolves raw and refined vegetable oils as well as light and medium stand oils. It is not, however, a good solvent for highly poly-merized stand oils and blown oils. In this case the addition of a proportion of a stronger solvent such as an aromatic or dipentene is necessary to prevent separation on standing.

Generally white spirit is a poor solvent for resins, and its use as the sole solvent in oil/resin combinations such as oleoresinous varnishes or alkyd resins is limited to the long-oil types of media. As the proportion of oil decreases, stronger solvents have to be introduced, and the short-oil types of alkyd will tolerate little, if any, white spirit. Comparatively few resins are soluble in white spirit; examples that are soluble are rosin, ester gum, soft and modified phenolic resins, and Damar resin. White spirit also dissolves asphaltic bitumen and petroleum pitch.

In addition to the extensive use of white spirit as a paint thinner, large quantities are consumed in the manufacture of polishes, and for cleaning purposes.

Flash point of commercial white spirit: 41°C (105°F). TLV 175 mg/m³.

Distillate

This can be regarded as a type of white spirit containing a higher proportion of high-boiling hydrocarbons. The boiling range is about 150° to 275°C (302° to 527°F) and consequently the material contains an appreciable *tail*. When used in paints this tail holds the film open much longer than normal white spirit does, and distillate is sometimes used to lengthen the wet-edge time of brushing finishes and undercoats.

VM & P naphtha (Varnish Makers' and Painters' naphtha)

This is a petroleum fraction marketed in the United States. It is somewhat similar to white spirit.

Solvents from coal tar

The complete change from coal gas to North Sea gas in the U.K. has resulted in the closure of a high proportion of the gasworks and a corresponding fall in the production of coal tar. Although some coal-tar solvents continue to be produced, the bulk of the solvents hitherto obtained from coal tar are now supplied by the petroleum industry.

Aromatic hydrocarbons

Toluol (commercial toluene), $C_6H_5CH_3$

This is the most extensively used diluent for cellulose nitrate lacquers. It is not a solvent for cellulose nitrate, but solutions in true solvents will tolerate greater additions of toluol than of other hydrocarbons. This has an important bearing on the cost of the lacquer. Toluol is a solvent for a large number of resins and is miscible with drying oils and most other solvents. It does not dissolve PVC, copals, or shellac.

The range of commercially available toluols is described in BS 805 (Toluenes):1977.

Flash point 4°C (39·2°F). Class 1. Highly flammable liquid.
TLV 375 mg/m³; 100 ppm.

Xylol (commercial xylene)

This is a mixture of the three isomers in which the meta-isomer predominates:

o-xylene
b.p. 144·2°C (291·6°F)

m-xylene
b.p. 139·1°C (282·4°F)

p-xylene
b.p. 138·3°C (280·9°F)

In some grades small quantities of other aromatic solvents are also present.

Xylol is now one of the most important solvents used in the paint industry. This is the result of its high solvent power for a very wide range of resins used in both stoving and rapid air-drying coatings, together with a convenient rate of evaporation. The solvent power is increased appreciably on addition of 10 to 20 percent n-butanol, and this mixture is very widely used in stoving finishes. The rate of evaporation is too great for use in brushing finishes.

Xylol is used as a diluent in cellulose nitrate lacquers and as the main solvent in lacquers based on polystyrene, polymethylmethacrylate and chlorinated rubber. It is also an excellent solvent for asphaltic bitumen and petroleum pitch.

In BS 458:1977 (Xylenes) there are three principal grades based on boiling range, i.e. the difference in temperature between 5 percent and 95 percent distillation. Thus for 3° xylene the range is 3°C lying between 137·5° and 144·5°C. For 5° and 10° xylenes the boiling ranges lie between 137° and 145·5°C, and between 135° and 148°C, respectively. The specific gravity range for 3° xylene is 0·860 to 0·875, but in this and the other grades the specification includes low-gravity material.

The flash point of 3° xylene is 24·4°C (76°F), but mixtures with n-butanol (flash point 36·7°C, 98°F) can flash below 22·8°C (73°F) and are then subject to the restrictions on highly flammable solvents laid down in the Petroleum Act. Xylol is classed as a highly flammable liquid.

TLV: 435 mg/m^3; 100 ppm.

High aromatic solvents from petroleum

These range in specific gravity from 0·833 to 0·895 and contain high percentages (80 to 93) of aromatics, principally the isomers of trimethyl benzene. They are supplied in graded boiling ranges between 100° and 210°C (212° and 410°F) and have flash points above 32°C (89°F).

They possess high solvent power and the odour is faint and not unpleasant. These solvents are marketed under proprietary names [14] and have largely replaced the coal-tar solvents.

TLV (trimethyl benzenes): 120 mg/m^3; 25 ppm.

Solvent naphtha

A solvent which is sometimes referred to as 90/160 naphtha from the fact that 90 percent distils below 160°C (320°F). It is used as a solvent in bituminous paints and in anti-fouling compositions.

Heavy naphtha

A strong-smelling solvent, often referred to as 90/190 naphtha, with excellent solvent power. In spite of the latter property, heavy naphtha has been largely replaced by the high aromatic solvents (*above*).

Terpene solvents

Turpentines

These solvents are obtained from the resinous materials present in pine and other coniferous trees. If a cut is made in the trunk of one of these trees a gummy material is exuded and can be collected by cups attached to the trunk. Distillation of this material yields the volatile "gum" turpentine. Steam distillation of the wood or of the stumps of pine trees yields "wood" or "stump" turpentine.

Another variety known as sulphate turpentine is obtained as a by-product in the wood-pulp industry and is obtained chiefly from Sweden.

For many years turpentine was the standard thinner for paints and varnishes, but since the introduction of alkyds and other synthetic resins together with supply difficulties during World War II, it has been replaced by other solvents.

The composition of turpentine is related to that of dipentene and pine oil and for this reason a survey of its properties is included.

Composition and properties

The turpentines consist of mixtures of members of the terpene group of unsaturated cyclic hydrocarbons. The mixtures consist chiefly of α-pinene, β-pinene, carene, sylvestrene, and dipentene and the properties vary with the source of the turpentine. α-pinene is the main constituent of gum turpentine, whilst the wood turpentines consist mainly of dipentene.

The turpentines are clear colourless liquids with a characteristic pungent odour. Long exposure to the vapour can cause headache and nausea.

They evaporate uniformly but, due to the unsaturation, a degree of oxidation and polymerization takes place at the same time. The product (linalool) so formed is non-volatile but miscible with drying oils and resins and therefore becomes part of the film.

The physical characteristics of the turpentines are as follows:

	Gum turpentine	*Wood turpentine*
Specific gravity	0·862–0·872	0·859–0·875
Distillation range	150°–180°C	150°–170°C
	(302°–356°F)	(302°–338°F)
Refractive index ($D_{20°C}$)	1·469–1·478	1·463–1·483
Flash point	30°–37°C	34°–36°C
	(86°–99°F)	(93°–96°F)
Non-polymerizable material	Not more than 11%	Not more than 16%

The turpentines are good solvents, and the Kauri–butanol value ranges from 50 to 90. In this respect wood turpentine, which consists largely of dipentene with small amounts of pinene, is a better solvent than the gum variety.

Dipentene

This solvent is prepared by a modification of the distillation process for wood turpentine. It has a higher distillation range (175° to 195°C, 347° to 383°F) and an appreciably lower rate of evaporation. It is an excellent solvent for many synthetic resins and is often used to improve the wet-edge time of decorative undercoats and finishes. It also possesses some anti-skinning properties. The physical characteristics of dipentene are:

Specific gravity	0·844–0·850
Boiling range	175°–195°C (347°–383°F)
Refractive index ($D_{20°C}$)	1·472
Flash point	54°C (130°F)
Kauri–butanol value	105

Dipentene forms coloured complexes with some metals, and the green coloration resulting from its reaction with cobalt is sometimes apparent in white enamels.

Pine oil

Pine oil is usually obtained with turpentine and is separated by fractional distillation. The chief constituent is terpineol. It has a higher boiling range (200° to 230°C, 392° to 446°F) than turpentine and a very much lower rate of evaporation. The specific gravity ranges from 0·932 to 0·945 and the refractive index ($D_{20°C}$) from 1·475 to 1·483.

Pine oil has an agreeable odour and is used as a re-odorant in paints and disinfectants. It also possesses some germicidal properties.

Alcohols

Methyl alcohol (methanol), CH_3OH

Large quantities of methyl alcohol are produced by reacting carbon monoxide and hydrogen at high pressures and in presence of a catalyst. The reaction temperature is about 300°C (572°F).

$$CO + 2H_2 \rightarrow CH_3OH$$

The mixture of carbon monoxide and hydrogen is obtained from methane present in petroleum gases. The methane is reacted with oxygen at high temperature (about 1400°C, 2552°F):

$$CH_4 + \tfrac{1}{2}O_2 \rightarrow CO + 2H_2$$

An alternative method utilizes a high-temperature reaction between carbon dioxide and hydrogen:

$$CO_2 + 3H_2 \rightarrow CH_3OH + H_2O$$

In both these processes the methyl alcohol is condensed out of the system.

Methanol is readily oxidized to formaldehyde and this constitutes one of its principal uses. It has a limited use in varnishes and stains. The physical properties are:

Specific gravity (20°C)	0·792
Boiling range	95% between 64° and 67°C
	(147° and 153°F)
Flash point	6°C (42°F)

Methyl alcohol has pronounced toxic properties and is used as a denaturant for the ethyl alcohol in methylated spirit. It is a cumulative poison and can lead to giddiness, unconsciousness and in extreme cases to impaired sight or permanent blindness.

Ethyl alcohol (ethanol), C_2H_5OH

Two general methods are used for the production of ethyl alcohol, the choice being determined by economic considerations.

Where there is an ample supply of fermentable material such as potato starch or molasses the fermentation process is employed. Enzymes contained in malt convert the starch into sugars, and these are converted to ethyl alcohol and carbon dioxide by the yeast enzymes. The alcohol is removed from the final liquor by distillation.

The second general method uses ethylene as the raw material and is widely used in the U.S.A. There are two methods by which ethylene can be converted to ethyl alcohol.

In the older method ethylene is dissolved in 95 percent sulphuric acid under pressure and forms a mixture of mono- and diethyl sulphates. On hydrolysis, ethyl alcohol is produced and is separated by distillation.

The more modern method is to pass a mixture of ethylene and steam over a catalyst at 300°C (572°F) and at high pressure. The reaction is a reversible one:

$$CH_2 = CH_2 + H_2O \rightleftharpoons C_2H_5OH$$

and the alcohol is condensed out of the system.

Properties and uses

Pure anhydrous ethyl alcohol has the following characteristics:

Specific gravity (20°C)	0·7937
Refractive index	1·3619
Boiling point	78·3°C (172·9°F)
Flash point	14°C (57°F)

Anhydrous alcohol is not usually encountered outside the research laboratory. The material used as a solvent is known as Industrial Methylated Spirit (IMS) and consists of ethyl alcohol (95 percent by volume) denatured with 5 percent methyl alcohol. Grades of IMS are available ranging from 61° overproof (91·84 percent by volume) to 74° overproof (99·24 percent by volume).

The sale and use of Industrial Methylated Spirit is very tightly controlled by the Customs and Excise Authorities.

Ethyl alcohol forms azeotropic (constant boiling point) mixtures with a number of other solvents and with water. The mixture with water boils at 78·15°C (172·67°F) and contains 95·57 percent ethyl alcohol and 4·43 percent water.

Industrial Methylated Spirit dissolves a number of natural resins including shellac, with which it forms several spirit varnishes. It also dissolves some synthetic resins such as polyvinyl acetate and cyclohexanone resin. It is completely miscible with water, hydrocarbons and castor oil.

n-propyl alcohol (n-propanol), $CH_3CH_2CH_2OH$

This is obtained as a by-product in the American Hydrocol process for production of hydrocarbons by the interaction of carbon monoxide and hydrogen in presence of a catalyst. It can also be obtained by the hydration of propylene.

The physical constants of the pure alcohol are:

Specific gravity (20°C)	0·8052
Refractive index	1·3860
Boiling point	97·2°C (207°F)
Flash point	22°C (71·6°F)

It forms azeotropic (constant boiling point) mixtures with a number of other solvents and with water. The latter mixture boils at 87·7°C (189·9°F) and contains 71·7 percent n-propyl alcohol and 28·3 percent water.

n-propanol dissolves rosin, shellac, ester gum and castor oil. It is sometimes used as a substitute for ethyl alcohol.

Iso-propyl alcohol (isopropanol), $(CH_3)_2CHOH$

Isopropanol is prepared by the hydration of propylene, $CH_3CH=CH_2$. The latter is dissolved in strong sulphuric acid under pressure to give a mixture of isopropyl sulphates. The mixture is then hydrolysed to give isopropyl alcohol. The overall reaction can be represented by the equation

$$\begin{array}{c} CH_3 \\ \diagdown \\ CH \\ \diagup\diagup \\ CH_2 \end{array} + \begin{array}{c} OH \\ | \\ H \end{array} \longrightarrow \begin{array}{c} CH_3 \\ \diagdown \\ CHOH \\ \diagup \\ CH_3 \end{array}$$

The crude alcohol obtained in this way is purified by azeotropic distillation. Pure isopropyl alcohol has the following physical constants:

Specific gravity (20°C)	0·7925
	(commercial isopropanol 0·787)
Refractive index	1·3776
Boiling point	82·4°C (180·3°F)
Flash point	12°C (53°F)

Isopropanol has a somewhat stronger odour than ethyl alcohol and, like the latter, is miscible with water in all proportions. It is very similar to ethyl alcohol in solvent properties and has been used extensively as a substitute for ethanol. It is not a solvent for cellulose esters but possesses latent solvent properties, in that small additions increase the solvent power of the true solvents. In admixture with small quantities of esters it dissolves cellulose nitrate.

n-butyl alcohol (n-butanol), $CH_3(CH_2)_2CH_2OH$

The older process of manufacture is dependent upon the fermentation of maize starch and molasses by a special enzyme. Some acetone and other alcohols are produced at the same time. Distillation then yields a product containing about 98 percent *n*-butyl alcohol, the balance being small amounts of acetone and other (chiefly ethyl) alcohol.

There are two modern processes for the production of *n*-butanol. One of these is based on the "Oxo" process [15] in which carbon monoxide and hydrogen are added to olefines in the presence of a cobalt catalyst to give aldehydes. Under these conditions propylene yields *n*-butyraldehyde together with some isobutyraldehyde:

$$CH_3CH=CH_2 + CO + H_2 \rightarrow CH_3CH_2CH_2CHO$$

$$(+ \text{ some } CH_3CH(CH_3)CHO)$$

High-pressure hydrogenation of the aldehydes, using a cobalt catalyst, then yields the alcohol:

$$CH_3CH_2CH_2CHO + H_2 \rightarrow CH_3CH_2CH_2CH_2OH$$
$$(+ some\ CH_3CH(CH_3)CH_2OH)$$

The *n*-butanol is then purified by distillation.

The alcohol can also be obtained by hydrogenation of crotonaldehyde, which is itself prepared from acetaldehyde. The reaction is carried out at 180°C (356°F) using a nickel-chromium catalyst:

$$CH_3CH = CH.CHO + 2H_2 \rightarrow CH_3CH_2CH_2CH_2OH$$

n-butanol possesses the following physical constants:

Specific gravity (20°C)	0·811
Boiling point	118°C (244·4°F)
Refractive index	1·3992
Flash point	37°C (98°F)

It dissolves a number of natural resins and the following synthetics: urea-formaldehyde, melamine-formaldehyde, polyvinyl acetate. It is miscible with hydrocarbon solvents and linseed oil.

n-butanol is used extensively in conjunction with xylol in stoving finishes, where it appears to act as a very efficient latent solvent. It is also used as a damping agent for cellulose nitrate and is used in high-flash and brushing lacquers where it also increases the blush resistance.

Secondary butyl alcohol (methylethylcarbinol), $CH_3CHOH.C_2H_5$

This is prepared from butylene, $C_2H_5CH = CH_2$ by a method similar to that used for isopropyl alcohol (*q.v.*). It has the following physical constants:

Specific gravity (20°C)	0·806
Refractive index	1·397
Boiling point	99·0°C (210°F)
Flash point	21°C (70°F)

It is extensively used as a medium boiling-point solvent in lacquers and is a latent solvent for cellulose nitrate.

Diacetone alcohol, $CH_3COCH_2C(OH)(CH_3)_2$

A colourless, odourless liquid with the following physical properties:

Specific gravity (20°C)	0·942
Refractive index	1·424
Boiling point	167°C (333°F)
Flash point	67°C (153°F)

Diacetone alcohol tends to decompose slowly on heating or long storage, especially in the presence of alkali. The main product is acetone. It is a good solvent for cellulose nitrate and is a useful high-boiling solvent for cellulose acetate lacquers.

Esters

Esters are produced by the general reaction between an alcohol and an acid according to the equation

$$R.COOH + R'.OH \rightleftharpoons R.COOR' + H_2O$$

The reaction is a reversible one and ultimately a state of equilibrium is reached. If however the water is removed by the use of a small quantity of concentrated sulphuric acid or other powerful dehydrating agent, the esterification will proceed to completion.

Some esters can be synthesized from unsaturated hydrocarbons in petroleum gases; for example, ethyl acetate can be synthesized according to the following scheme:

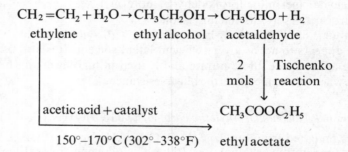

$$CH_2=CH_2 + H_2O \rightarrow CH_3CH_2OH \rightarrow CH_3CHO + H_2$$

ethylene   ethyl alcohol   acetaldehyde

2 mols | Tischenko reaction

acetic acid + catalyst $CH_3COOC_2H_5$

150°–170°C (302°–338°F)   ethyl acetate

The esters prepared from the lower alcohols are volatile liquids, some of which are used in cellulose ester lacquers. They have high solvent power and the different esters are generally similar in properties, the main difference being in the distillation range and rate of evaporation. For a given alcohol, the boiling point of the ester increases with increasing size and complexity of the molecule of acid used. The same general rule holds for a given acid with increase in the alcohol molecule.

The lower esters possess characteristic fruity smells which in some cases can be very pungent. Their main use is in cellulose ester lacquers, and the following esters are among the most widely used.

Ethyl acetate, $CH_3COOC_2H_5$

A low boiling solvent of high solvent power, which is now being replaced gradually by methylethyl ketone which is appreciably cheaper. The British Standard Specification [16] for ethyl acetate specifies specific gravity (20°C, 68°F) 0·901 to 0·904. Boiling range—a minimum of 95 percent between 76° and 78°C (169° and 172°F).

n-butyl acetate, $CH_3COOC_4H_9$

This is the most widely used medium-boiling solvent for cellulose nitrate lacquers. It is sufficiently volatile to leave the film readily but does not cause chilling of the film with consequent blushing. It is an excellent solvent for many other resins and oils but is not used to any degree outside cellulose nitrate lacquers. In these it is gradually being replaced by methyl isobutyl ketone which is cheaper and possesses greater solvent power. The specific gravity at 20°C (68°F) is 0·879 to 0·882 and the boiling range from 124° to 128°C (255° to 262°F).

Ethyl lactate, $CH_3.CH(OH)COOC_2H_5$

This is a very efficient high-boiling solvent for cellulose ester lacquers. It has high solvent power for other resins and will tolerate the addition of considerable amounts of diluent. It imparts good flow to the lacquer and high gloss to the film, but the slow evaporation rate causes the film to remain soft for some time.

Ethyl lactate is also used in small quantities in stoving lacquers where it assists *secondary flow* in the stove and eliminates any film imperfections. The specific gravity at 20°C (68°F) is 1·032 to 1·038; boiling range, 95 percent between 135° and 160°C (275° and 320°F).

Ethers

Ethers are produced by the elimination of a molecule of water from two molecules of alcohol by the use of a powerful dehydrating agent such as concentrated sulphuric acid. The two alcohol molecules may be identical or different.

$$R.OH + HO.R' \rightarrow R—O—R' + H_2O$$

The lowest member of the group is dimethyl ether, $CH_3.O.CH_3$ which is a gas. *Diethyl ether*, $C_2H_5.O.C_2H_5$ (ordinary ether) is a very volatile liquid with a boiling point of 34·5°C (94·1°F), and is highly flammable. It is used in laboratories as an extraction solvent.

As the size of the alkyl group or groups increases, the ethers become less volatile. They are generally inert (apart from flammability) toward the majority of chemicals and are useful solvents for oils and a number of natural resins.

Di-isopropyl ether, $(CH_3)_2CH.O.CH(CH_3)_2$

Used industrially for low-temperature and other extractions. It is less volatile than diethyl ether and is a solvent for rosin and ester gum, oils, hydrocarbons and rubber. It is not a solvent for synthetic resins.

Di-isopropyl ether is a dangerous material as it forms very unstable peroxides on exposure to the air and these explode on heating. Their formation can be minimized by storage in the dark and addition of reducing agents such as hydroquinone or catechol or by storage over 20 percent sodium hydroxide solution.

The formation of unstable peroxides on exposure to light and air is a property of most of the lower ethers.

Glycol ethers

A series of very useful ethers is produced from the dihydric alcohol ethylene glycol, $HO.CH_2.CH_2OH$. This substance can form mono- or di-ethers according to the number of hydroxyl groups reacted, but the mono-ethers are the more important. The free hydroxyl groups in the mono-ethers can be further reacted with acids to produce ether-esters (*below*).

The glycol ethers possess mild, pleasant odours and the following are the most important members of the series. They are marketed in this country under the proprietary names "Oxitol" (Shell Chemical Company) and "Cellosolve" (Union Carbide).

Ethylene glycol monomethyl ether, $HO.CH_2CH_2.OCH_3$, 2-*methoxyethanol*

("Methyl Oxitol", "Methyl Cellosolve"). As the lowest member of the series this has the lowest boiling point and highest evaporation rate. The specific gravity is 0·96 to 0·97, boiling range 123° to 125·5°C (253° to 258°F) and flash point 38·9°C (102°F). It is a solvent for both cellulose acetate and cellulose nitrate.

Ethylene glycol monoethyl ether, $HO.CH_2CH_2OC_2H_5$, 2-*ethoxyethanol*

("Oxitol", "Cellosolve"). This is a slightly hygroscopic liquid with specific gravity 0·930 to 0·931, boiling range 133·5° to 135·5°C (272·3° to 275·9°F) and flash point 42·8°C (109°F). It is a good solvent for cellulose nitrate and ethyl cellulose and has a high degree of tolerance for aromatic diluents. It has a low rate of evaporation and is also a solvent for many natural resins and synthetics such as alkyds and epoxies.

Ethylene glycol monobutyl ether, $HO.CH_2CH_2OC_4H_9$, 2-*butoxyethanol*

("Butyl Oxitol", "Butyl Cellosolve"). A hygroscopic liquid with specific gravity 0·902, boiling point 171·2°C (341°F), and flash point (Pensky-Martens) 67·8°C (154°F). A good solvent for cellulose nitrate with an evaporation rate lower than that of the ethyl ether. It therefore gives good flow and reduced tendency to "orange peel". It also has a high tolerance for aliphatic diluents and is used in brushing lacquers.

Glycol ether–esters

The free hydroxyl group in the glycol mono-ethers can be esterified to produce a useful range of ether–esters of which the following material is typical.

Ethylene glycol monoethyl ether acetate, $CH_3CO.O.CH_2CH_2.O.C_2H_5$,
    2-ethoxyethyl acetate

("Oxitol acetate", "Cellosolve acetate"). A colourless liquid with an ester odour. Specific gravity 0·972 to 0·975; boiling range 145° to 165°C (293° to 329°F); flash point (Pensky–Martens) 52·2°C (126°F). A good high-boiling solvent for cellulose nitrate, ethyl cellulose, and many natural resins. It is also a good solvent for many synthetic resins.

Ketones

Ketones are a class of compound characterized by the presence of the carbonyl group $\rangle C = O$. The carbon can be attached to two alkyl groups which are the same, as in acetone, $CH_3CO.CH_3$, or different as in methyl ethyl ketone $CH_3COC_2H_5$. Alternatively the carbonyl group can form part of a ring, such as cyclohexanone (*see below*). Other reactive groups such as hydroxyl may be present in the molecule as in diacetone alcohol,

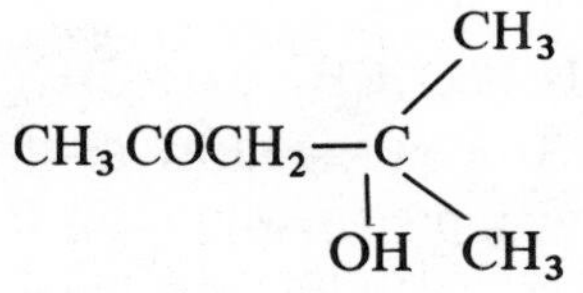

$$CH_3\,COCH_2\!-\!\underset{\underset{\textstyle OH}{\displaystyle |}}{C}\!\!\underset{\textstyle CH_3}{\overset{\textstyle CH_3}{<}}$$

The lower members of the class are very volatile and highly flammable liquids and are soluble in water. With increasing size of substituent groups the members lose the water-solubility and become less volatile and less flammable. The highest members are solids.

The following ketones are among the most important paint and lacquer solvents.

Acetone (dimethyl ketone), CH_3COCH_3

This is prepared from isopropyl alcohol by passage over a copper catalyst at 500°C (932°F) and at three atmospheres pressure:

$$(CH_3)_2.CHOH \rightarrow CH_3COCH_3 + H_2$$

Acetone is an extremely volatile liquid with a flash point below 0°C. It therefore flashes at normal temperatures and very strict precautions are necessary in its use.

It is one of the most powerful organic solvents and is used as a low-boiling solvent in some cellulose acetate and nitrate lacquers. The great solvent power renders it an attractive ingredient for paint strippers.

Acetone possesses the following characteristics:

Specific gravity (20°C)	0·790
Refractive index ($D_{20°C}$)	1·3599
Boiling point	56·2°C (133°F)
Flash point	$-16·5°C$ (2°F)

Methyl ethyl ketone, $CH_3COC_2H_5$

Known commercially as MEK, this is prepared by the high-temperature dehydrogenation of secondary butyl alcohol:

$$CH_3 . CHOH . C_2H_5 \rightarrow CH_3CO . C_2H_5 + H_2$$

This solvent is similar to acetone in basic reactions but differs in the higher flash point and lower rate of evaporation. The physical characteristics are:

Specific gravity (20°C)	0·805
Refractive index ($D_{20°C}$)	1·379
Boiling point	79·6°C
Flash point	$-7°C$ (19°F)

MEK is used widely as a resin solvent.

Methyl iso-butyl ketone, $CH_3 . CO . CH_2 . CH . (CH_3)_2$

Also known as MIBK. It is prepared by the hydrogenation of mesityl oxide, an unsaturated ketone prepared by dehydrating diacetone alcohol. The two stages can be represented by the equations:

$$CH_3 . CO . CH_2 \underset{\underset{\text{diacetone alcohol}}{OH}}{C(CH_3)_2} \rightarrow \underset{\text{mesityl oxide}}{CH_3 CO . CH = C(CH_3)_2} + H_2O$$

$$\xrightarrow{+H_2}$$

$$CH_3 CO . CH_2 CH(CH_3)_2$$

MIBK possesses the following characteristics:

Specific gravity (20°C)	0·802
Refractive index ($D_{20°C}$)	1·396
Boiling range	95% between 114° and 117°C
	(237° and 243°F)
Flash point (Abel)	10°C (50°F)

It is a solvent for a very wide range of resins and is extensively used both in stoving enamels and lacquers where it functions as a medium boiling-point solvent.

Cyclohexanone

This is a ring compound with the structure

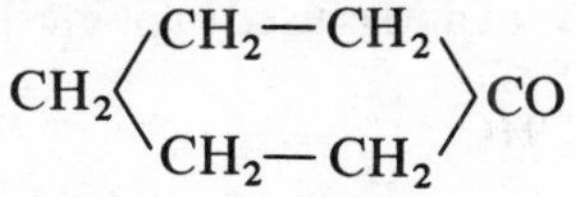

$$CH_2 \left\langle \begin{array}{c} CH_2 - CH_2 \\ CH_2 - CH_2 \end{array} \right\rangle CO$$

and can be produced by the dehydrogenation of cyclohexanol.

It has the following physical characteristics:

Specific gravity (20°C)	0·950
Refractive index	1·443–1·451
Boiling range	95% between 150° and 158°C
	(302° and 316°F)
Flash point	47°C (117°F)

It is an excellent solvent for a number of resins and is used in cellulose nitrate lacquers, in which it tolerates considerable additions of diluents such as toluene.

Chlorinated hydrocarbons

The hydrogen atoms in both the saturated and unsaturated hydrocarbons can be replaced by chlorine, and many of these substitution products are useful solvents.

As a class they possess the following characteristics:

(i) High specific gravity.

(ii) A low degree of flammability. The highly chlorinated members are completely non-flammable and whilst the lower chlorinated members can be ignited, the heat of combustion is so low that the flame is extinguished.

(iii) Characteristic odours.

(iv) Anaesthetic and often pronounced toxic properties.

(v) A low heat of vaporization, indicating high volatility in relation to the boiling point.

(vi) High solvent power for a wide range of organic resins and coatings. It is unfortunate that the desirable properties of non- or low-flammability and high solvent power are accompanied by toxic properties. The range of application is therefore severely limited, and whenever they are used special precautions must be taken to protect operators from the toxic effects.

The following members of the class are encountered in the surface coatings industry.

Methylene chloride, CH_2Cl_2

A highly volatile liquid with boiling point 41°C (106°F) and specific gravity 1·336. It is a very powerful solvent and is the major component of many non-flammable paint strippers.

Ethylene dichloride, $ClCH_2 . CH_2Cl$ (1:2-*dichloroethane*)

Boils at 84°C (183°F) and has a specific gravity of 1·44. Like methylene chloride it is a powerful solvent and is used in paint strippers.

Trichloroethylene, $Cl_2C = CHCl$

Boils at 87°C (189°F) and has a specific gravity of 1·48. It is a powerful solvent and evaporates very rapidly. Among its many uses are extraction of oils from crushed seeds, dry-cleaning of fabrics, and recently it has been used successfully as the sole solvent in a dipping paint.

Perchlorethylene, $Cl_2C = C . Cl_2$

This is very similar to trichlorethylene in properties except for the evaporation rate which is lower.

PLASTICIZERS

Plasticizers are used in lacquers or non-convertible coatings which dry by solvent evaporation only, and in some types of stoving finish. Their function is to improve flexibility and extensibility of the film without impairing other properties. They should, ideally, be non-volatile, completely compatible with other film components, and chemically stable. The important classes of plasticizers are the following.

Esters

These comprise the largest class. The boiling points of esters increase with increasing molecular weight until the volatility of the material decreases to the point where it can be used as a plasticizer. Esters with boiling points above 250°C (482°F) have, generally, a very low vapour pressure at normal temperatures.

The following list comprises a few of the many ester plasticizers in common use, with their boiling points and the materials for which they are satisfactory plasticizers:

Plasticizer	Boiling point (at 760 mm) (101·325 kN/m²)	Plasticizer for:
Triacetin	280°C (536°F)	cellulose acetate cellulose nitrate
Dimethyl phthalate	284°C (543°F)	cellulose acetate
Dibutyl phthalate	340°C (644°F)	cellulose nitrate ethyl cellulose vinyls
Dioctyl phthalate	230°C (445°F) at 5 mmHg (666·61 Nm²)	cellulose nitrate, vinyls
Tributyl phosphate	290°C (554°F)	cellulose nitrate
Tritolyl (tricresyl) phosphate	420°C (788°F)	vinyls, cellulose nitrate

Plasticizer	Boiling point (at 760 mm) ($101 \cdot 325$ kN/m^2)	Plasticizer for:
Butyl stearate	365°C (689°F)	cellulose nitrate
Dibutyl sebacate	345°C (653°F)	cellulose nitrate
Tributyl citrate	225°C (437°F) at 5 mmHg ($666 \cdot 61$ Nm^2)	cellulose nitrate
Butyl oleate	203°C (397°F) at 10 mmHg ($1333 \cdot 22$ Nm^2)	chlorinated rubber ethyl cellulose

The properties of a range of interesting trialkyl phosphate plasticizers have been described by Quinn [17].

Camphor

Camphor was the first plasticizer to be incorporated into cellulose nitrate, the combination being known as celluloid. It is a ring ketone of the terpene family and possesses the structure:

$$
\begin{array}{ccc}
CH_2 & \rule{1cm}{0.4pt} & CH_2 \\
| & CH_3 & | \\
| & | & | \\
CH & \rule{0.3cm}{0.4pt}\ C\ \rule{0.3cm}{0.4pt} & CH \\
| & | & | \\
| & CH_3 & | \\
CH_2 & \rule{1cm}{0.4pt} & C{=}O
\end{array}
$$

Both natural and synthetic types of camphor are available and are used in cellulose nitrate lacquers.

Camphor is a solid with specific gravity $0 \cdot 992$ and melting point 178°C (352°F). It has an appreciable vapour pressure at room temperatures, so it is very gradually lost from the film.

Castor oil

The *first pressings* castor oil (Chapter 11) is used in cellulose nitrate lacquers where permanent flexibility is required, as in leather finishes. It is a non-drying oil, completely non-volatile, and remains permanently in the film.

Blown castor oil

This is a higher-viscosity material produced by blowing air through the raw oil, and is used as a plasticizer in cellulose undercoats and primers.

Chlorinated waxes

A range of liquid plasticizers derived from hydrocarbon waxes is marketed under the trade name "Cereclor" (ICI Ltd.). They are completely non-volatile and are used as plasticizers for chlorinated rubber, vinyls, and cellulose nitrate.

References to Chapter 9

[1] SCATCHARD, G., *Chem. Rev.*, 1931, **8**, 321.
[2] HILDEBRAND, J. H. & SCOTT, R. L., *The Solubility of Non-electrolytes*, 3rd edn. Reinhold, 1950.
[3] BURRELL, H., *Off. Dig.*, 1955, **27**, No. 369, 726.
[4] *Idem*, The challenge of the solubility parameter concept. *J. Paint Tech.*, 1968, **40**, No. 520, 197.
[5] HANSEN, C. J., The three-dimensional solubility parameter. (1) Key to paint component affinities. *J. Paint Tech.*, 1967, **39**, No. 511, 505.
HANSEN, C. J. & SKAARUP, K. (2) Independent calculation of the parameter components. *Ibid.*, 511.
[6] CROWLEY, J. D., TEAGUE, G. S. & LOW, J. W., A three-dimensional approach to solubility. Part 1, *J. Paint Tech.*, 1966, **38**, No. 496, 269; Part 2, *ibid.*, 1967, **39**, No. 504, 19.
[7] HOY, K. L., New values of the solubility parameters from vapor pressure data. *J. Paint Tech.*, 1970, **42**, No. 541, 76.
[8] TEAS, J. P., Graphic analysis of resin solubilities. *J. Paint Tech.*, 1968, **40**, No. 516, 19.
[9] NELSON, R. C., HEMWALL, R. W. & EDWARDS, G. D., Treatment of hydrogen bonding in predicting miscibility. *J. Paint Tech.*, 1970, **42**, 636.
[10] RUDD & TYSALL, *JOCCA*, 1949, **32**, 546.
ELVEN, RUDD & TYSALL, *JOCCA*, 1950, **33**, 520.
LOIBLE & STOLTON, FATIPEC Congress, 1962, p. 228.
[11] BS 245:1976. Specification for mineral spirits (white spirit and related hydrocarbon solvents) for paints and other purposes. British Standards Institution, London.
[12] Threshold Limit Values (TLV) published by the Health and Safety Executive and taken from the American Conference of Governmental Industrial Hygienists (ACGIH).
[13] BS 2511:1970. Methods for the determination of water (Karl Fischer method). British Standards Institution.
[14] "Aromasol" (ICI); "Shellsol" (Shell Chemicals); "Solvesso" (Esso Petroleum).
[15] WADDAMS, A. L., *Chemicals from Petroleum*, 2nd edn. John Murray, London, 1977.
[16] BS 553:1965. Ethyl acetate. British Standards Institution.
[17] QUINN, J. A. W., Some newly available trialkyl phosphates. *POCJ*, 1970, p. 949.

10 Drying oils, driers and drying

The oils used in the paint industry are derived mainly from vegetable and, to a much lesser extent, from animal sources. They are esters of glycerol and fatty acids, non-volatile, and unstable at high temperatures.

These oils vary in properties according to the nature of the fatty acids combined with the glycerol, i.e. they may be saturated or unsaturated. In the latter case, the oil possesses the valuable property of setting slowly to a solid and adherent film when spread on a surface and exposed to the air. This film formation is an irreversible process and the film is insoluble in white spirit. This process is known as "drying" and the oils can be classified into the following groups based on their drying properties:

> Drying oils, e.g. linseed.
> Semi-drying oils, e.g. soya bean, tobacco seed.
> Non-drying, e.g. castor.

CONSTITUTION OF DRYING OILS

The esters produced by combination of glycerol with fatty acids are known as glycerides. A number of fatty acids are involved, so that the number of possible glycerides is large. Most oils consist of glycerides of several acids in varying proportions.

Glycerol (the pure material is known as glycerine) is a trihydric alcohol derived from propane. It is a colourless, viscous liquid, very hygroscopic and possesses the following constants: boiling point, 290°C (554°F); specific gravity (15°C) 1·265; refractive index 1·473.

As a trihydric alcohol, glycerol is capable of combining with one, two or three acid radicals to form mono-, di- and triglycerides. The constitution of these can be represented as follows, where R is a fatty acid chain:

$$
\begin{array}{llll}
CH_2OH & CH_2OCOR & CH_2OCOR & CH_2OCOR \\
| & | & | & | \\
CHOH & CHOH & CHOH & CHOCOR'' \\
| & | & | & | \\
CH_2OH & CH_2OH & CH_2OCOR' & CH_2OCOR' \\
\text{glycerol} & \text{monoglyceride} & \text{diglyceride} & \text{triglyceride}
\end{array}
$$

Vegetable and animal oils consist mainly of the triglycerides with only small proportions of the mono- and diglycerides.

171

Fatty acids are a large group of compounds consisting of long hydrocarbon chains of about 16 carbon atoms and upwards, attached to a carboxyl (–COOH) group. They are usually straight chain compounds and are of two types (i) *saturated*, in which the four valencies of each of the carbon atoms are fully satisfied, and (ii) *unsaturated*, which contain one, two or three ethylenic linkages or double bonds. The two types can be represented thus:

Saturated acid — no double bonds
$$CH_3CH_2CH_2(CH_2)_{14}CH_2.COOH$$

Unsaturated acid
$$CH_3CH_2CH=CH.CH_2CH=CH.CH_2CH=CH(CH_2)_7.COOH$$

In unsaturated acids containing more than one double bond, the latter may be isolated from each other as in the example preceding, or they may be grouped as closely as possible together and separated by only one C—C linkage. In this case they are said to be "conjugated". The following is an acid of this type:

$$CH_3(CH_2)_3.CH=CH.CH=CH.CH=CH.(CH_2)_7.COOH$$

Conjugated unsaturated acid (elaeostearic)

Natural glycerides contain both saturated and unsaturated fatty acids, but it is to the latter that the drying properties are due. The types of fatty acids present in the common drying oils are shown in Table 10.1.

Furthermore, the reactivity and drying properties of conjugated unsaturated acid glycerides is much greater than that of compounds containing isolated double bonds. This accounts for the reactivity of tung oil which contains the conjugated unsaturated elaeostearic acid depicted above.

Drying oils of the linseed type consist of glycerides of fatty acids containing two or three isolated double bonds, whilst semi-drying oils contain acids with only one or two double bonds. This class includes soya bean and tobacco seed oils. Non-drying oils, e.g. castor, consist of glycerides from saturated fatty acids which have no drying properties, or may contain small amounts of acids with one double bond.

The natural vegetable oils rarely consist of pure glycerides, but each glycerol molecule is combined with two or three different acids which may be saturated or unsaturated. The possible combinations are large, and as the esters are not particularly stable it is possible that the changes in properties which occur during storage or treatment are due to a rearrangement of acid radicals between the various glycerides.

Again, partial decomposition of glycerides during ripening or processing can lead to the liberation of free fatty acid. The presence and amount of this acid can influence the behaviour of the oil.

Table 10.1—Fatty acids

Acid	Molecular formula	No. of double bonds	Occurrence
Myristic	$C_{14}H_{28}O_2$	None	
Palmitic	$C_{16}H_{32}O_2$	,,	Non-drying oils. Small quantities occur in a number of drying and semi-drying oils.
Stearic	$C_{18}H_{36}O_2$	,,	
Arachidic	$C_{20}H_{40}O_2$	,,	
Oleic	$C_{18}H_{34}O_2$	1	Majority of vegetable oils
Erucic	$C_{18}H_{34}O_2$	1	Rape oil
Ricinoleic	$C_{18}H_{34}O_3$	1	Castor oil
Linoleic	$C_{18}H_{32}O_2$	2	Most vegetable oils. High percentage in poppy seed, safflower, soya bean, sunflower and tobacco seed.
Linolenic	$C_{18}H_{30}O_2$	3	Linseed, perilla and stillingia oil
Elaeostearic	$C_{18}H_{30}O_2$	3	Tung oil
Licanic	$C_{18}H_{28}O_3$	3	Oiticica oil
Clupanodonic	$C_{22}H_{34}O_2$	4	Small quantity in fish oils

The constitutions of the most commonly encountered unsaturated fatty acids are as follows:

Oleic. $CH_3(CH_2)_7CH=CH.(CH_2)_7.COOH$
Linoleic. $CH_3(CH_2)_4CH=CH.CH_2.CH=CH.(CH_2)_7.COOH$
Linolenic. $CH_3CH_2CH=CH.CH_2.CH=CH.CH_2.CH=$
$$CH.(CH_2)_7.COOH$$

Elaeostearic. $CH_3(CH_2)_3.(CH=CH)_3.(CH_2)_7.COOH$
Licanic. $CH_3(CH_2)_3.(CH=CH)_3(CH_2)_4.CO.(CH_2)_2.COOH$

CHEMICAL EXAMINATION OF DRYING OILS

Acid value

The amount of free organic acid in an oil is important in many applications and is expressed as the number of milligrams of potassium hydroxide required to neutralize the acids in one gram of the oil. It is determined by dissolving a known weight of oil in a neutral alcohol/benzene solution and titrating with 0·1 M sodium or potassium hydroxide, using phenolphthalein as indicator. The method is described in detail in BS 242 [1].

Saponification value

When a glyceride is treated with sodium or potassium hydroxide it is saponified or decomposed into the alkali salts of the fatty acids and glycerol. The saponification value is defined as the number of milligrams of potassium hydroxide required to saponify one gram of oil. It is determined by boiling a known weight of the oil with an excess of 0·5 M alcoholic potassium hydroxide. The excess alkali is then titrated with 0·5 M hydrochloric acid, a blank on the alcoholic potash solution being run at the same time. The method is described in detail in BS 242.

Unsaponifiable matter

Drying oils contain small quantities of materials other than esters and these are not saponified on boiling with alkali. One percent is the normal maximum, and figures above this generally denote impurity. In the determination of unsaponifiable matter, the oil is saponified with alkali and the product extracted with ether, in which the soaps are insoluble. The ether extract is evaporated to dryness and the residue weighed.

Iodine value

The double bonds in an unsaturated glyceride react with iodine to form saturated compounds. The weight of iodine (in grams) required per 100 grams of oil is known as the "iodine value". It is a direct measure of the degree of unsaturation and therefore of the drying properties of the oil. The method introduced by Wijs is normally employed in this country and is described in detail in BS 242. Briefly, the method involves dissolving a known weight of the oil in carbon tetrachloride and then adding an excess of iodine solution (in the form of iodine monochloride). The mixture is kept in the dark for one hour to allow complete reaction and the excess of iodine is estimated by titration with standard sodium thiosulphate solution. A blank experiment on the iodine solution is carried out at the same time.

A low iodine value indicates a poor quality oil or adulteration with a non-drying oil. With oils of the linseed type, containing isolated double bonds, the iodine value gives a correct indication of the degree of unsaturation. With conjugated double bonds, however, addition takes place at points × :

$$—CH=CH.CH=CH—+I_2 \rightarrow —CHI.CH=CH.CHI—$$
$$\times \times$$

and the iodine value obtained is therefore lower than the true value.

In the United States the Hanus method is employed which uses iodine monobromide.

Insoluble bromide test

When bromine is added to unsaturated fatty acids containing three or more isolated double bonds the resulting bromides are insoluble in ether. The bromides from fatty acids containing one or two double bonds are soluble in ether. Conjugated double bonds also give soluble bromides. The insoluble bromides can be separated, dried and weighed. The result is expressed as a percentage increase in the weight of the oil, and the following figures are indicative of the values obtained: linseed 34 to 36, soya bean 4, wood oil nil. The test can be used to differentiate between vegetable and fish oils.

Maleic or Diene value

This value indicates the content of acids containing a conjugated double bond. It is based on the Diels–Alder reaction in which conjugated double bonds react with maleic anhydride in the following manner:

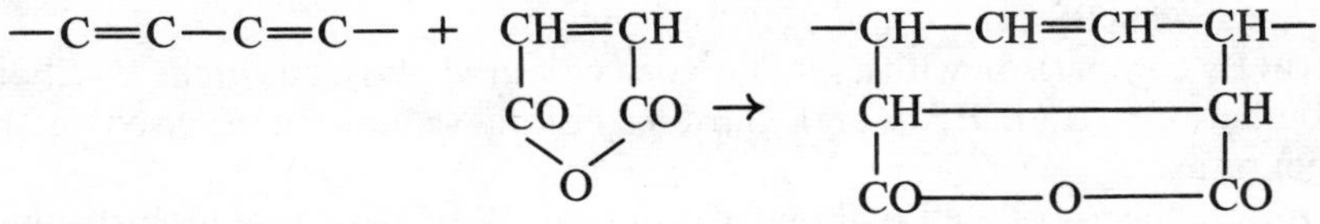

$$
-C{=}C{-}C{=}C- \;+\; \begin{array}{c} CH{=}CH \\ | \quad | \\ CO \quad CO \\ \diagdown \diagup \\ O \end{array} \rightarrow \begin{array}{c} -CH{-}CH{=}CH{-}CH- \\ | \qquad\qquad | \\ CH{-}\!\!-\!\!-\!\!-\!\!-\!\!-CH \\ | \qquad\qquad | \\ CO{-}\!\!-\!\!O\!\!-\!\!-CO \end{array}
$$

In the determination, the oil is refluxed with excess maleic anhydride in toluene and the residual maleic anhydride estimated. This is done by conversion to maleic acid by addition of water followed by titration with standard sodium hydroxide solution. A blank determination is done on the maleic anhydride solution, and hence the weight combined with the oil can be calculated.

Acetyl value

If uncombined hydroxyl groups are present, as in castor oil or oils containing mono- or diglycerides, they can be reacted with acetic anhydride or acetic acid to give acetyl derivatives. These acetyl derivatives can be saponified to yield the original compound and acetic acid or its salt. The acetyl value is defined as the number of milligrams of potassium hydroxide required to neutralize the acetic acid produced by saponification of one gram of the acetyl compound.

Hydroxyl value

This is closely related to the acetyl value and is defined as the number of milligrams of potassium hydroxide equivalent to the acetic acid which combined with 1 gram of the hydroxyl-containing oil.

Flash point

The flash points of vegetable oils fall generally within the range 210° to 290°C (400° to 550°F), but the figures for individual oils can vary to some extent with variations in composition, quality and source of the seed, and method of extraction. It can also be affected by the acid value.

The determination of the flash point is carried out in the Pensky–Martens apparatus; figures appreciably lower than those quoted previously are usually an indication of contamination.

PHYSICAL PROPERTIES OF DRYING OILS

Colour

The colour of an oil is of importance as it can influence the colour of the varnish or paint in which it is incorporated. There are three methods by which the colour can be assessed:

(i) By comparison with a series of solutions of iodine in potassium iodide as specified in BS 242.

(ii) By comparison with a set of graded coloured glasses as in the Lovibond Comparator. Both PRA and Gardner colour scales can be used in this apparatus.

(iii) By the use of the Lovibond Tintometer, BDH pattern. This instrument is specially designed to take cells containing liquids. White light is passed through the cell to one half of the eyepiece. White light from an identical source is passed through a box containing graded coloured glass filters until a match is obtained. The operation of the glass filters is identical with that in the Lovibond–Schofield Tintometer (Chapter 2).

Specific gravity

The specific gravities of the drying oils all fall within a narrow range, generally between 0·925 and 0·933. The bodied oils are a little higher. The specific gravity is not therefore of great value in identification, although it is very useful in formulations involving the oils. It is determined by the usual methods using hydrometer or specific gravity bottle.

Refractive index

The refractive power of an oil is of importance in respect of its relationship to the refractive index of pigments which are dispersed in it. The opacity of a pigment, and therefore of the paint, is a function of the difference between the refractive indices of pigment and oil or other medium.

The refractive index of an oil is best determined on a standard instrument such as the Abbe refractometer in which the angle of total light reflectance from the surface is measured directly. This instrument is described in most physics textbooks.

The range of refractive indices of drying oils is comparatively narrow, from 1·46 to 1·54, but the difference between individual oils is sufficient for use as a test for purity and as an aid to identification.

Viscosity

All liquids flow when a force is applied, some more readily than others. Thin limpid liquids like water or white spirit are said to possess a low viscosity, whilst high viscosity is a characteristic of stand oils. Viscosity can be regarded as resistance to flow and can be defined more precisely as follows.

Suppose two planes, each 1 sq. cm. in area, are immersed in a liquid parallel to each other and 1 cm apart. If the force required to move one plane in a direction parallel to the other at a velocity of 1 cm per second is 1 dyne, then the liquid has a viscosity of 1 poise (after Poiseuille).

On this basis, distilled water has a viscosity of 0·01005 poise or 1·005 centipoise at 20°C. This is used as a standard.

Measurement of viscosity

There are four general methods in use for the measurement of the viscosity of oils by which they can be compared either directly or indirectly with distilled water. These comprise the U-tube (Ostwald), falling-sphere and efflux viscometers, and the bubble tubes. The general principles on which these viscometers operate are as follows.

U-tube—measurement of rate of flow of a definite volume of liquid through a capillary of definite bore. The instrument is illustrated diagramatically in Fig. 10.1. The liquid is introduced by a pipette into the right-hand limb of the instrument and is sucked up into the left-hand limb so that it occupies the section from just above A to just above mark C. The viscometer is suspended in a thermostatically controlled bath and when the correct temperature has been attained, the liquid is allowed to flow back into bulb D, and the time taken for it to fall from A to B is measured.

This viscometer is used for the comparison of the viscosities of two liquids and the results are treated in the following way:

Let η_1 and η_2 be the viscosities of the two liquids
Let ρ_1 and ρ_2 be the densities of the two liquids
Let t_1 and t_2 be the times taken from A to B by the two liquids.
Then $\dfrac{\eta_1}{\eta_2} = \dfrac{(\rho_1 t_1)}{(\rho_2 t_2)}$

Hence if one of the viscosities is known, e.g. if distilled water is first used in the apparatus, the viscosity of the second liquid can be calculated from this equation.

This method is described in detail in BS 188:1977 [2].

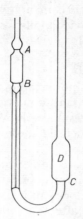 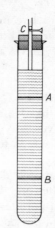

Fig. 10.1 — U-tube viscometer   Fig. 10.2   Falling-sphere viscometer

Falling sphere — measurement of the time taken by a metal sphere to fall through a column of liquid of definite length.

The apparatus used consists of a glass tube, 2 to 3 inches in diameter and as long as possible consistent with immersion in a thermostatically controlled bath. Theoretically the tube should be as wide as possible to eliminate wall effects, but in practice volume considerations are important as the liquid must acquire the thermostat temperature in a limited time.

The tube (Fig. 10.2) has two widely separated marks A and B and is suspended in a thermostat at 25°C (77°F). The metal sphere C of known radius (about 1 mm) is allowed to fall between the two marks and the time measured.

The principle of the method is illustrated by the following form of the Stokes equation:

$$\eta = \frac{2r^2(D-d)g}{9v}$$

where η = viscosity of the liquid
D = density of sphere
d = density of liquid
g = gravitational constant
v = terminal velocity of sphere.

If S be the distance between the marks and t the time of fall of the sphere, the above equation can be written in the form:

$$\eta = \frac{2}{9}r^2g\left(\frac{D-d}{S}\right)t$$

If a number of spheres of known radius are used and r^2 is plotted against

$1/t$ the result should be a straight line from which η can be obtained. Alternatively, the method can be used for direct comparison of the viscosity of a liquid with that of distilled water. The method is described in detail in BS 188:1977.

Efflux viscometer [3]. This usually consists of a metal cup with parallel sides and with an accurately machined orifice in the centre of the base. The volume is generally 100 cc. The time for the cup to empty through the orifice is noted and the results expressed in seconds. These cups are unsuited to paints which possess any degree of thixotropy, and since a high proportion of paints possess some structure the use of cups has been largely superseded by that of viscometers in which the structure is broken down before the viscosity is measured. Efflux cups continue to be used as control instruments in industrial application of paints which possess near-Newtonian flow properties.

Bubble tubes. If a small glass tube, e.g. a test-tube, is almost filled with liquid and a cork inserted leaving an air bubble, on inversion of the tube the bubble will rise at a rate depending on the viscosity of the liquid. The tube can be compared with a graded series of standard viscosity oils sealed in tubes of similar length and diameter. The viscosity of the liquid can then be recorded in terms of the nearest standard tube(s). This method is not highly accurate but is useful for works control.

Sets of standards are produced by the Paint Research Association, Teddington, and numbered 1 to 20. The American Gardner–Holt tubes are based on the same principle and designated by letters A to Z.

Viscosity of oils

The majority of drying and semi-drying oils possess viscosities similar to linseed oil (about 40 centipoises), but tung oil is somewhat higher. The viscosities of stand and blown oils are very much higher, and these are generally designated by viscosity, e.g. 30-poise stand oil.

Viscosity of oil mixtures

When two oils of differing viscosity are mixed in equal proportions, the viscosity of the mixture is given approximately by the relationship:

$$\log \eta = \tfrac{1}{2} (\log \eta_1 + \log \eta_2)$$

where η is the viscosity of the mixture, η_1 and η_2 the viscosities of the individual oils.

When x parts of oil of viscosity η_1 are mixed with y parts of oil of viscosity η_2, then the viscosity of the mixture is given by:

$$(x + y) \log \eta = x \log \eta_1 + y \log \eta_2$$

The accuracy of this expression depends on the difference between η_1 and

η_2. If these are widely different the actual viscosity of the mixture is less than that calculated.

Odour

Drying oils possess characteristic odours which are more pronounced in the raw than in the refined oils. Odour is a guide to identification but is of little value in the assessment of purity, unless the adulterant is present in appreciable amount and also possesses a definite odour.

THE DRYING PROCESS

When a film of drying oil is exposed to the air it sets gradually, and ultimately "dries" to a tough elastic film. Whereas the original oil is soluble in solvents such as white spirit or dipentene, the dried film is not soluble in these solvents and is therefore quite different in nature from the original oil. It is this drying process which enables one coat of paint to be applied over another.

The chemical reactions involved in the drying process are very complex, but a major part is played by atmospheric oxygen and reactions at the double bonds of the unsaturated glycerides. It is thought that after an initial induction period, oxygen is absorbed on the carbon atoms adjacent to the ethylenic groups to form hydroperoxides of the general type $-CH(OOH).CH=CH-$. These hydroperoxides then undergo a series of reactions resulting in the formation of a number of compounds including short-chain carboxylic acids. In addition, free radicals are formed and in their brief existence they induce polymerization by cross-linking between double bonds.

The processes involved in drying are markedly affected by temperature. It is probable that polymerization plays a greater part at high than at low temperatures. In the latter the main process is one of oxidation. The whole drying process takes place much more slowly in the absence of light, so the reactions are, in part at least, photo-catalysed. Drying is also retarded by traces of sulphur dioxide and other impurities in the air [4].

Straight drying oils, such as linseed, dry very slowly, some two to three days being required for raw linseed oil. The process can, however, be accelerated very considerably by the introduction of certain metals in the form of oil-soluble compounds or soaps. The effective metals are lead, cobalt and manganese and, to a lesser extent, cerium, vanadium and zirconium. They are metals which can exert more than one valency, and their catalytic activity is doubtless associated with this fact.

When lead compounds alone are present, the setting of the film is accelerated but the surface remains tacky for a considerable time. Lead appears to act as a polymerization catalyst or "through" drier.

Cobalt acts as an oxidation catalyst and, when used alone, causes rapid drying of the surface, the underfilm remaining soft. As a result "rivelling" of the surface takes place. The use of a mixture of lead and cobalt in certain proportions gives uniform surface and through drying, and the effect is greater than would be expected from the individual effects.

The catalytic effect of drier metals is proportional to the quantity added within fairly narrow limits. With lead these limits are 0·5 to 1·0 percent as metal calculated on the oil, whilst with cobalt the limits are 0·05 to 0·1 percent. The figures for manganese, the drying action of which appears to be between those of lead and cobalt, are similar to those for cobalt. No increase in rate of drying is obtained by increasing the quantity of driers above the figures quoted. To do so can have an adverse effect, the film becoming sticky or "sweating back".

The mechanism of drying of oleoresinous varnishes and of resins containing fatty acids derived from drying oils (alkyds, epoxy esters and urethane oils) is, in general, similar to that of oils. There is, however, an important difference. The presence of other ingredients results in faster drying times and harder films. The quantities and relative amounts of drier metals are adjusted to suit the amounts and nature of the fatty acids present: the quantities used are appreciably less than those used for oils.

Paints for children's toys and for decorative work generally are now made free from lead, and drying is achieved by using a mixture of cobalt and zirconium driers (*below*).

The action of drier metals does not cease when the film is dry. It continues throughout the life of the film and so contributes to the ultimate embrittlement and breakdown.

The drying effect of the metal compounds appears to be independent of the type of acid radical, provided the solubility in paint media is complete.

TYPES OF DRIER COMPOUNDS

Naphthenates. "Naphthenic acid" is the term given to a mixture of acids of high molecular weight obtained from petroleum crudes. The acid value of the material is generally between 170 and 230 and the drier metal salts are supplied to the industry as solutions in petroleum solvents. The solutions contain a specified metal content. The usual concentrations are:

Lead naphthenate:	24% metallic lead
Cobalt naphthenate:	6% metallic cobalt
Manganese naphthenate:	6% metallic manganese

but other concentrations are available as well as mixed solutions.

Naphthenate driers can be incorporated into paint or varnish by simple stirring or can be added to varnish during the cooling-down stage. They possess a characteristic and strong smell and this can persist for some time after the paint is applied. For this reason the octoate driers (*see below*) are preferred for inside decorative work.

Zirconium naphthenate, in admixture with the cobalt compound, has been found to be an effective replacement for lead naphthenate where lead-free paints are required.

Cerium naphthenate has been used to some extent as a replacement for lead naphthenate. Like the latter it is used in conjunction with manganese or cobalt, but the action differs somewhat from that of lead in that the film remains soft for a period and then hardens rapidly.

Vanadium naphthenate behaves in a similar manner to cobalt but it has a darkening action on the oil or varnish.

Calcium naphthenate is often added to alkyd gloss finishes in addition to other driers (lead/cobalt or cobalt/zirconium). It does not function as a drier but is thought to confer the following advantages: (a) prevention of precipitation of drier metals (notably lead) by certain acidic components of the medium, and (b) reduction of the tendency of "crystalline bloom" to form on the surface of the paint film.

Octoates (ethylhexoates) are made from the colourless and fairly pure octoic (ethylhexoic) acid. Generally they are superior to the naphthenates in respect of colour, possess little or no smell, and permit the production of low-viscosity solutions of higher metal content than the naphthenates. Common concentrations of metal are:

Lead octoate:	33% metallic lead
Cobalt octoate:	12% metallic cobalt
Manganese octoate:	12% metallic manganese.

There is virtually no residual smell after application of paints containing this drier.

Liquid driers. The type encountered consists usually of a solution of lead and cobalt naphthenates in white spirit.

Terebine driers. This term is sometimes used as an alternative for liquid driers but is also applied to a solution containing a dark boiled oil containing lead and manganese in white spirit.

References to Chapter 10

[1] BS 242 (inc. 243, 259 & 632): 1969. Linseed oil. British Standards Institution, 2 Park Street, London, W1A 2BS.
[2] BS 188:1977. Methods for the determination of the viscosity of liquids. British Standards Institution.
[3] BS 3900: Methods of test for paints. Part A6:1971. Determination of flow time.
[4] HOLBROW, G. L., Atmospheric pollution: its measurement and some effects on paint. *JOCCA*, 1962, **45**, 701.

11 Natural and modified drying oils

Extraction of oils

The oil-bearing seed is cleaned and freed as completely as possible from foreign matter and other seeds by screening and winnowing. It is then subjected to the following processes.

Crushing

The seeds are ground to a meal by passage over a vertical three- or five-roll mill. The pressure applied, and therefore the fineness of the meal, is adjusted to the type of seed being processed. If the meal is too fine, subsequent operations are more difficult.

Heating

The meal is then heated to 80° to 90°C (176° to 194°F) with live steam. This serves to disrupt the cells and facilitates removal of the oil, which is separated from the meal by pressing, expelling, or extracting.

Pressing

A measured quantity of the meal is wrapped in filter cloth and placed in an open hydraulic press. The oil is expelled on application of pressure. This type of press has been found satisfactory for linseed, but for meals with a higher oil content, e.g. castor, the cage press is generally used. This consists of a cylinder with perforated walls which is filled with the meal. Application of pressure to the end of the cylinder expels the oil through the perforated walls. This is essentially a batch process.

Expelling

This is carried out in a screw press comprising a worm shaft rotating in a perforated cylinder. The diameter of the cylinder decreases toward the outlet so that the meal is subjected to steadily increasing pressure. The oil is expelled through the perforations and the extracted meal is discharged at the end. Expelling is a continuous process and is the most widely used extraction method. It is well suited to extraction of meal from nuts, for example, in the preparation of tung oil.

Solvent extraction

For this purpose the meal is dried and rolled into flakes. The flakes are then treated with the solvent (petroleum ether or a chlorinated hydro-carbon) in a closed vessel and allowed to stand for some time at a temperature below the boiling point of the solvent. The time allowed is generally about twenty minutes. The solution of oil in solvent is then drained off and the process repeated two or three times.

The counter-current method of extraction is often used. In this, fresh meal meets solvent which has been used in previous extractions, whilst the almost exhausted meal is treated with clean solvent. In this way maximum use is made of solvent and the meal is completely extracted.

The solvent is then distilled off and the oil stored in tanks. The process gives a higher yield than pressing or expelling, but the oil has a higher acid value and is darker in colour. In view of the cost of solvents and solvent losses in the process, an efficient solvent recovery process is essential if solvent extraction is to be economical.

LINSEED OIL

Source

Linseed is the seed of the flax plant which is grown in many parts of the world, but the relative values of the flax and seed depend on climate and local conditions. The chief exporting countries are the Argentine and India. A high-quality seed was formerly produced in the Baltic countries but this is no longer available.

Processing

After the seed has been reduced to meal it is heated and the oil is extracted by the impellor method. The yield of oil obtained is about 30 percent of the weight of the seed, but the total oil content of the latter is about 35 percent. The oil so obtained is a thin, brownish-yellow liquid known as raw linseed oil.

Raw linseed oil

The expressed oil has a very characteristic odour, but this, in common with certain other properties, can vary with the source of the seed, the weather conditions during growth, and the degree of maturity when harvested.

The fatty acids present are chiefly linolenic and linoleic with a small quantity of saturated fatty acids. The following figures give the approximate composition of oil from Argentine seed: linolenic 37·5%, linoleic 31·8%, oleic 16·5%, saturated acids 9%, glycerol 4·4%, unsaponifiables 0·8%.

The physical constants of raw linseed oil are set out in Table 11.1. Raw linseed oil is used in the manufacture of glazing putty, and sometimes in oil paints, e.g. red lead. The oils more often used in oil paints are the refined varieties (*below*).

Refined linseed oils

Heat refining

If raw linseed oil is heated slowly, a light gelatinous material separates at between 200° and 250°C (392° and 482°F). This insoluble matter is known as "break" and consists mainly of phosphatides (see Lecithin, p. 191). Removal of this break is essential if the oil is to be used in any process involving heat treatment. Removal is effected by a filter press designed to collect the filtrate. Oil which has been purified by removal of break in this way is known as heat-refined linseed oil, but the term "raw" is occasionally — but incorrectly — applied.

Alkali refining

Raw linseed oil generally contains a small amount of free acid (Table 11.1). If the oil is agitated and heated with slightly more sodium hydroxide than is required to neutralize the acids and then allowed to stand, an aqueous layer separates. This contains sodium salts of the fatty acids as well as the break and any water from the oil. It is run off and acidified to give *linseed acid oil*.

The treated linseed oil is washed thoroughly with water and then dried by vacuum or heat. The product is known as alkali-refined or varnish linseed oil (VLO) and the physical constants are given in Table 11.1.

Table 11.1 — Characteristics of linseed oils

	Raw	Acid-refined	Alkali-refined	Boiled	Stand oils	Blown oils
Specific gravity (15·5°C)	0·931–0·936	0·930–0·934	0·930–0·934	0·94–0·95	0·95–0·98	0·96–1·0
Viscosity	40 cp.	approx. 40 cp.	approx. 40 cp.	0·8–1·2 poise	3 poise upwards	3 poise upwards
Refractive index (20°C)	1·4800–1·4835	1·4810–1·4825	1·4810–1·4825		1·481–1·490	
Acid value	1–3	7–8	0–0·5	2–4	4–12	4–6
Saponification value	190–195	190–195	190–195			
Iodine value	175–185	175–185	175–185	160–170	130–150	100–140
Unsaponifiable matter	0·7–1·0%		1–2%			
Drying* time	4 days	4 days	4 days	16–24 hours	24–36 hours	24–36 hours

*Without driers

Acid refining

The raw oil is placed in a lead-lined vessel and fairly concentrated sulphuric acid added in a fine stream or spray. About 1 percent on the weight of oil is generally used, and this destroys the break and much of the colouring-matter. The acid is then washed out with water and the oil dried by vacuum or heat. The oil is known as acid-refined oil and contains appreciable amounts of free fatty acids (acid value up to 8).

Acid-refined linseed oil is used widely as a grinding medium for pigments, the wetting process being facilitated by the fatty acid groups present. The physical constants are set out in Table 11.1.

Treated linseed oils

The double bonds in unsaturated glycerides will slowly absorb oxygen in the presence of air. The process takes place faster in the presence of drier metals and faster still if the oil is heated in the presence of air or driers. The process is mainly oxidation, but some polymerization takes place at the same time. These processes lead to an increase in viscosity and a decrease in the drying time of the oil.

Boiled oil

Raw (or heat-refined) linseed oil is heated in a vessel through which air is bubbled during the process. Heating is carried out by means of steam coils. Lead and manganese driers are added to the hot oil and the process becomes exothermic. The temperature is then controlled by passage of cold water through the coils. The heating is continued until the specific gravity reaches 0·942 at 15·6°C (60°F). A process of this type in which the temperature reaches a maximum of 137·5°C (279·5°F) produces *dark boiled oil*. If the temperature is kept lower and the manganese drier replaced by cobalt, a *pale boiled oil* results.

The formation of boiled oil is accompanied by an increase in viscosity and specific gravity but a decrease in drying time from 4 days to 12–20 hours. The physical properties are summarized in Table 11.1.

Pale boiled oil is frequently added to oleoresinous paints to improve flow and ease of brushing. It can also be added to alkyd-type undercoats for the same purpose.

Blown oil

This is a thickened linseed oil made by blowing air through the heated raw oil in the absence of driers. The oil is heated by steam coils to about 135°C (275°F), air being bubbled through. The reaction is exothermic and is controlled by passing water through the coils. The temperature is then maintained at 125° to 135°C (257° to 275°F) until the desired viscosity is obtained.

Blown oil is a brownish-yellow colour and yellows still further on exposure to light. The physical constants are set out in Table 11.1. Films produced from blown oil possess good initial gloss, but the gloss retention is poor. The higher viscosity grades give good flow and levelling properties to highly pigmented systems and have therefore been used in undercoats and flat wall paints.

The process used for blown oil preparation is the cheapest method of thickening linseed oil and consists mainly of oxidation.

Stand oil

When linseed oil is kept at a high temperature in the absence of driers it slowly thickens and the process can be used to prepare thickened oils of any desired viscosity by cooling at the appropriate moment. The thickening is essentially a polymerization process and the product is known as *stand oil*.

Alkali-refined oil (VLO) is used as the starting material and is heated to 280° to 300°C (536° to 572°F) until the desired viscosity is reached. This can take from six to eight hours. During the process a number of volatile decomposition products are formed and these can be removed by passing a current of air over the surface of the hot oil (giving *open pot* stand oil) or by passing an inert gas such as nitrogen over the oil. This gives *closed pot* stand oil which is paler than the open pot type and therefore in greater demand.

Stand oils are slow drying. They can be obtained in a wide range of viscosities, from 3 to 200 poise or more. Their basic physical constants are set out in Table 11.1. Stand oils are used as components of varnishes, both "cooked" and cold-cut, and to improve the levelling properties of undercoats based on alkyd or oleoresinous media.

Litho varnishes. These are very high-viscosity stand oils of the open pot type. A certain amount of oxidation has taken place in the manufacture, and such oils are admirably suited to lithographic printing.

Enamel oils. These are mixed stand oils, the best-known consisting of 3 parts linseed and 1 part tung oil. Other ratios, e.g. 9/1 and 5/1, are also used. They are prepared from mixtures of the raw oils in a manner somewhat similar to linseed stand oil, but the temperatures required are a little lower. The enamel oils give films that are harder and more water-resistant and alkali-resistant than the stand oils. They are often used in conjunction with phenolic resins to produce clear and pigmented marine coatings.

TUNG OIL

This oil is also known as "Chinese wood oil" or simply "wood oil" and is obtained from two species of the Aleurites family — *A. fordii* found originally in Central China and *A. montana* in southern China and Cambodia. The demand for the oil stimulated work on the propagation of the trees in other parts of the world and today *A. fordii* is being grown successfully in the United States and the Argentine and *A. montana* in Malawi. The former is in the more plentiful supply.

The nuts of these trees yield about 30 percent of a brownish-yellow oil with a characteristic odour. This raw tung oil is more viscous than other raw oils of the linseed type and differs appreciably from these in both composition and properties. The glycerides present in tung oil contain a high percentage of elaeostearic acid, a material containing three conjugated double bonds, and these confer a high degree of reactivity on the oil.

Raw tung oil contains no break, but the trielaeostearin can exist in two isomeric forms, designated α and β. The liquid α-form is the normal state, but under the influence of light it changes to the solid β-isomer. This product is not as easy to handle and is commercially less valuable. The conversion of the raw oil to a stand oil eliminates the tendency to form the β-isomer.

When a film of raw tung oil is exposed to the air it dries to a "frosted" finish which, under magnification, is seen to be a fine wrinkle. Use is made of this property in the formation of "wrinkle finishes", but for all other uses the raw tung oil is subjected to heat treatment.

Heat treatment

When tung oil is heated, the increase in viscosity is more rapid than with linseed oil, and if kept at about 280°C (536°F) it forms a stiff gel in a short time. This gelation is irreversible and the gell is insoluble in oils and solvents. If the gelation or polymerization is carried out by immersion of a test-tube (150 mm × 25 mm) containing a sample of the oil in a mineral oil bath at 280°C, the time taken for gelation is an indication of the purity of the tung oil. The end-point is best determined by placing a glass rod in the test-tube and raising it at intervals. When the oil gels, the test-tube and contents can be lifted by the rod. Under these conditions, *Aleurites fordii* oil gels in about 12 minutes but *A. montana* oil takes approximately double this time. Control of the temperature is important as gel times decrease with increasing temperature.

If tung oil is mixed with other oils the gelation time is increased and the gel is soft.

The polymerization reaction is exothermic and in the production of tung-oil stand oil rigid control of conditions is therefore essential. If the process gets out of hand, gelation takes place and there is a risk of fire as a result of increase in volume and overflowing of the pot. The oil must be bodied to the correct degree to eliminate the tendency to web or *gas check* when exposed to foul atmospheres from gas or coke flames. A correctly bodied oil is *gas-proof* and will dry to a hard brilliant film.

The tendency to gel on bodying can be checked by the addition of a highly acid resin such as rosin, and there is no risk of gelling when tung oil is bodied in presence of other oils such as linseed. The latter mixture is employed in enamel oils, which are made by heating linseed oil until the desired viscosity increase is obtained, adding the raw tung oil, and then heating at 260°C (500°F) to obtain the required viscosity.

Properties and uses

The physical constants of tung oil are set out in Table 11.2. It is widely used because of its rapid drying and high resistance to water penetration and saponification.

In combination with ester gum or modified phenolic resins it is used in a wide range of air-drying finishes. With 100 percent phenolic resins it provides marine spar varnishes, whilst with coumarone resin it produces a widely used medium for aluminium paints and for primers for alkaline surfaces such as plasters and cement render. In conjunction with phenolic resin it has been used to upgrade the chemical resistance of some alkyds.

DEHYDRATED CASTOR OIL

Raw castor oil is obtained from the seeds of the plant *Ricinus communis* which is widely distributed in tropical and semi-tropical countries. The oil is virtually colourless, non-drying, and has an abnormally high viscosity for a natural vegetable oil. It is soluble in alcohol and is used as a plasticizer and as a lubricant.

The principal acid present—ricinoleic—possesses one double bond and also a hydroxyl group in the molecule. If the oil is heated to about 260° to 280°C (500° to 536°F) in the presence of a catalyst, water is eliminated between the hydroxyl group and the hydrogen of a neighbouring carbon atom. This yields a mixture of two acids each containing two double bonds, but in one case they are conjugated:

$$CH_3(CH_2)_5CH(OH)CH_2CH = CH(CH_2)_7COOH$$
ricinoleic acid

$$CH_3(CH_2)_4CH = CH.CH_2.CH = CH(CH_2)_7COOH$$
9:12-octadecadienoic acid
(isolated double bonds)

$$CH_3(CH_2)_5CH = CH.CH = CH.(CH_2)_7.COOH$$
9:11-octadecadienoic acid
(conjugated double bonds)

The oil produced in this way is known as dehydrated castor oil (DCO), and by continuing the heat treatment, bodied oils of almost any desired viscosity can be obtained. The reaction is unique in one respect in that the viscosity

Table 11.2—Composition and properties of drying oils

	Special acids %	Linolenic acid %	Linoleic acid %	Oleic acid %	Saturated acids %	Iodine value %	Saponification value	Specific gravity	Acid value	Refractive index
Arachis (ground nut) – – – – –	–	–	13–26	51–71	14–22	85–89	188–196	0·916–0·920	4–10	1·470–1·472
Cashew nut – – – –	–	–	7·7	74–77	16·5–18	79–85	187–195	0·911–0·918	1·4–1·6	1·462–1·463
Castor – – – – –	Ricinoleic 80–85	–	3–4	7–9	10	82–90	176–187	0·958–0·969	1–4	1·477–1·479
Chia – – – – –	–	41	47	0·7	8·4	200	192	0·934	0·6	1·485
Corn (maize) – – –	–	–	39	43·4	8–10	103–125	188–193	0·921–0·928	4–8	1·467–1·474
Cotton seed – – –	–	–	30–39	33–50	20–25	103–113	190–193	0·921–0·932	0·5–4	1·460–1·475
Hemp – – – –	–	24	53	12·6	10	148–155	190–193	0·927–0·932	4	1·477–1·482
Linseed (raw) – –	–	37–42	31·7–37	9–16	6–9	175–185	190–195	0·931–0·936	1–3	1·480–1·484
Lumbang (candle nut) –	–	20·8	39·6	26·2	8·4	155–160	190–193	0·920–0·927	0·5–1	1·477–1·478
Niger seed – – –	–	–	54–57	31–33	10–17	126–134	189–198	0·924–0·928	8	1·467–1·469
Oiticica – – –	Licanic 73–78 Elaeostearic 4–9	–	–	4–6	11–13	179–218	188–192	0·966–0·969	3–15	1·514–1·516
Perilla – – – –	–	42·5	32–42	4–10	6–7	193–201	190–205	0·932–0·935	1–6	1·475–1·485
Poppy – – – –	–	–	62–65	25–30	7–8	133–169	190–198	0·924–0·927	0·7–3	1·475–1·478
Po-yok – – – –	Coupeic 41	–	–	9–10	12	140–157	188–192	0·955–0·965	1–17	1·502–1·516
Rubber seed – –	–	19·6	31·5	27·3	16	133–143	186–195	0·924–0·930	–	1·466–1·468
Safflower – – –	–	0·7–1	51–67	26–38	6–10	140–150	188–194	0·925–0·928	5–6	1·468–1·469
Soya – – – –	–	2–6	50–60	26–30	11–14	125–140	188–194	0·923–0·929	0·3–3	1·468–1·478
Stillingia – – –	–	8–11	52–58	25–26	6–9	160–180	204–210	0·936–0·946	3–10	1·482–1·483
Sunflower – – –	–	–	52–58·5	34–42	6–9	120–135	189–194	0·924–0·926	4–7	1·474–1·479
Tobacco seed – –	–	–	60–70	16–20	9–10	130–145	186–197	0·923–0·925	1–8	1·474–1·482
Tung – – – –	Elaeostearic 72–86	–	–	4–18	4–6	155–170	189–197	0·939–0·943	5–10	1·519–1·522
Walnut – – – –	–	10–16	73	17–29	5–9	132–160	190–196	0·925–0·927	–	1·469–1·471

of the oil actually falls during the first three hours from the original 7 to about $3\frac{1}{2}$ poise and then rises rapidly.

Properties and uses

The physical properties of DCO are set out in Table 11.2. The presence of an acid containing conjugated double bonds results in the oil resembling tung oil in some of its properties. Under certain conditions it will yield a frosted film on drying, and it will gel if heated for a sufficient time. The rates of bodying and gelling are not as great as those of tung oil.

When mixed with driers, DCO dries faster than linseed stand oils, to a tough film which, however, retains a characteristic *after-tack* for some time. It is therefore seldom used as an oil or varnish in air-drying paints, but in the form of varnishes it finds wide use in stoving finishes. It is also incorporated into air-drying and stoving alkyds in which it shows very good non-yellowing properties. In this respect it is far superior to linseed oil.

SEMI-DRYING OILS

Soya bean oil

The soya plant, *Glycine hispida*, is a native of China but is now cultivated on a large scale in the United States, Europe and elsewhere. The ground beans are extracted with solvent and yield 15 to 18 percent of oil, a high proportion of which is used for edible purposes. When intended for paint, the oil is treated to remove the lecithin.

Soya lecithin is a mixture of phosphatides and is marketed in the form of a semi-paste containing 30 to 40 percent of the parent oil. The lecithin is precipitated by addition of about 3 percent water to the warm soya oil, followed by thorough agitation. It is then centrifuged. Lecithin finds wide use in the paint industry as a surface-active agent. It is useful in pigment wetting and the control of flocculation.

Properties and uses of soya bean oil

The constitution and physical properties are set out in Table 11.2. As a semi-drying oil, soya is intrinsically slow drying and the product, after separation of the lecithin, takes about three times as long to dry as does linseed under the same conditions. However, the very small percentage of linolenic acid present results in good colour retention (little yellowing) and the films are far superior to linseed in this respect. The oil shows to greatest advantage when incorporated into alkyds. For this, the fatty acids are first separated and then incorporated into the resin. These resins are frequently described as *linoleic rich*.

Soya modified alkyds are now widely used for non-yellowing white paints and in these the slow-drying characteristics of the fatty acids appear to have been lost. These alkyds are described in more detail in Chapter 13.

Tobacco seed oil

The tobacco plant, *Nicotiana tabacum*, grows in many parts of the world and supplies of oil to this country originate mainly in India. The oil is extracted by pressure or solvent and has a yellowish-brown colour and a characteristic odour. It is bleached before processing for use in the paint industry.

Properties and uses

The constitution and physical properties are set out in Table 11.2. It is a typical semi-drying oil and takes much longer to dry than does linseed under similar conditions. The film remains soft and somewhat tacky but is non-yellowing as a result of the absence of linolenic acid from the triglycerides.

Tobacco seed oil is marketed as bleached alkali-refined oil and as stand oil, both of which are non-yellowing and slow drying. The bulk of the oil, however, is treated to separate the fatty acids and these are then incorporated into alkyds. These alkyds dry satisfactorily and are valued for their non-yellowing properties and durability. They are described in more detail in Chapter 13.

The following semi-drying oils are generally similar to tobacco seed. Their characteristics are set out in Table 11.2.

Safflower seed oil. The flower is grown in India and Russia and the seeds yield an oil with a high linoleic acid content. The fatty acids are separated and used to produce alkyds with good non-yellowing properties.

Sunflower seed oil is obtained from the seeds of the variety *Helianthus annuus* and is pale yellow. It is mainly used in non-yellowing alkyds.

Poppy seed oil is obtained from the seeds of the opium poppy, *Papaver somniferum*, and is used mainly as a medium for artists' oil colours. The properties of pale colour, slow drying, and good colour retention make it a very suitable medium for artists.

Fish oils

The oils derived from various species of fish differ considerably in their drying properties. Those obtained from fish such as the pilchard and herring dry extremely slowly, but the North American menhaden and sardine give oils with fair drying properties. They all possess very powerful fishy odours.

The poor drying properties result from the presence of appreciable amounts of saturated acids and the fact that the unsaturated acids present dry only slowly. Thus menhaden oil, which is typical of the class, possesses the following composition and properties:

Glycerides of unsaturated acids	72–78%
Glycerides of saturated acids	24–28%
Unsaponifiable matter	1%

Specific gravity	0·929–0·933
Iodine value	150–180
Saponification value	189–193
Acid value	4–7
Refractive index	1·474–1·482

The saturated acids present are those normally found in vegetable oils — myristic, palmitic and stearic, but the unsaturated acids are very different. About 20 percent of oleic acid is present, together with a number of other unsaturated acids, some of which contain 4 to 6 isolated double bonds.

The proportion of unsaturated acids can be increased by a process known as "segregation"; these acids can be used to replace the vegetable oil fatty acids, in whole or in part, in some types of alkyd and urethane oil.

Tall oil

This is obtained as a by-product in the production of wood pulp by the Kraft process which involves treatment of the coniferous woods with sodium sulphite and caustic soda. In the purification of the wood pulp a residue or "black liquor" is obtained which contains sodium soaps. This is purified and then acidified to give a dark-coloured crude tall oil. The following figures have been quoted for the composition and properties:

Fatty acids	41–59%
Rosin acids	32–48%
Volatile and unsaponifiable matter	10%
Saponification value	169
Acid value	160

The colour is improved by a decolorizing process or by distillation. The rosin acids can be crystallized out at lower temperatures, and as the product is virtually a mixture of acids it can be converted to salts of drier metals (tallates) or incorporated into alkyds.

MODIFIED DRYING OILS

A number of methods have been suggested for improving the properties of drying oils, and the most interesting of these involve the fatty acids. The methods in most general use are the following.

Segregation. The oil is treated with a solvent with which it is not completely miscible, and allowed to stand. Two layers are formed which differ in the ratio of oil to solvent. In addition, a partial separation of saturated from unsaturated glycerides takes place. This separation is reflected in the drying times, one fraction drying more rapidly and the other more slowly than the parent oil. By repeating the process on the fractions a further segregation of glycerides can be effected.

This process has been used to improve the drying properties of soya bean oil by treatment with furfural, whilst liquid propane has been used in the treatment of pilchard oil.

Reaction with unsaturated compounds. The unsaturated compounds used are those which react with the double bonds in the unsaturated fatty acid chains. The most widely used compounds are maleic anhydride, cyclopentadiene and styrene.

Compounds with maleic anhydride. This material reacts with both conjugated and isolated double bonds in the following manner:

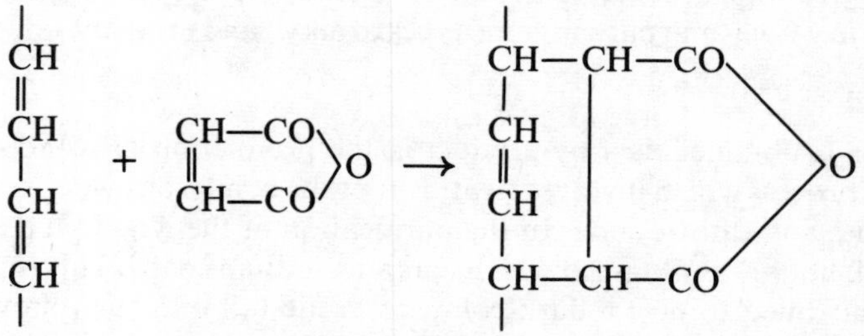

Reaction with conjugated double bonds

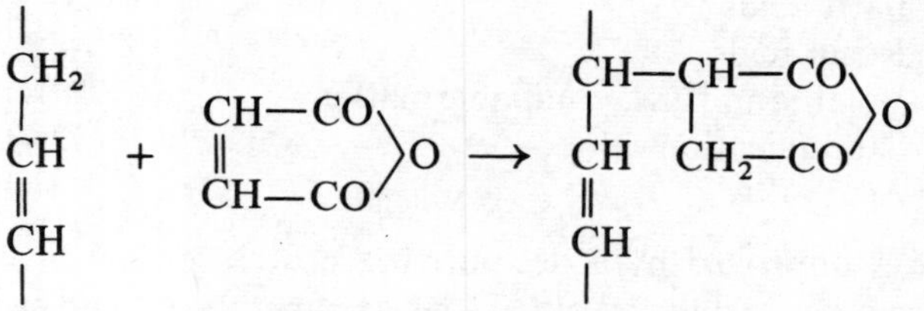

Reaction with isolated double bond

The products can be further esterified with glycerol or pentaerythritol to give the so-called maleinized oils. These oils body faster than the parent oils and give films with improved hardness, water resistance and colour.

Water-soluble oils. When an oil such as linseed is reacted with maleic anhydride and the product neutralized with ammonia, the resulting oil is soluble in water. Such materials are of interest for *water-soluble* paints.

Compounds with cyclopentadiene. Cyclopentadiene reacts with unsaturated drying oils when they are heated together under pressure to produce oils which dry rapidly to a high gloss. A range of cyclopentadienized oils is possible with varying viscosities and drying rates depending on the proportion of cyclopentadiene employed. Cyclopentadiene is assigned the following structural formula:

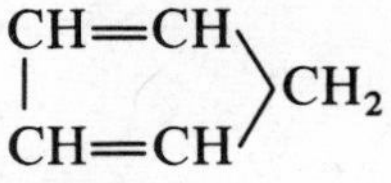

$$\left.\begin{array}{c} CH\!=\!CH \\ | \\ CH\!=\!CH \end{array}\right\rangle CH_2 \quad \text{\textbf{Cyclopentadiene}}$$

Compounds with styrene. Styrene or vinyl benzene, $C_6H_5CH\!=\!CH_2$, can be heated with unsaturated oils to give interesting products. The properties of these styrenated oils depend on the proportion of styrene used. They dry quickly to hard films but are not widely used in the paint industry. The reaction is used, however, on a considerable scale for the production of styrenated alkyds (Chapter 13).

Mono- and diglycerides of fatty acids. These are not normally found in natural drying oils but can be prepared fairly readily from these materials. The oil is heated to 270° to 280°C (518° to 536°F) with glycerol and a catalyst (litharge or alkali) in an inert atmosphere. The product is a mixture consisting mainly of mono- and diglycerides together with a little unchanged triglyceride. Separation can be effected by extraction with methyl or ethyl alcohol.

The monoglycerides are the more important. They are occasionally used as plasticizers for spirit varnishes but their main use is in the manufacture of alkyds. The two free hydroxyl groups are very readily esterified.

12   Natural and modified natural resins

Natural resins are obtained from many varieties of trees which exude a viscous liquid when the bark is damaged. This exuded *oleoresin* then hardens and acts as a protective for the damaged area. With the exception of rosin which is derived from pines, natural resins are obtained from tropical forest trees.

These resins are insoluble in water but dissolve to varying degrees in organic solvents. This property serves to distinguish them from the natural *gums* which, although obtained from certain species of trees, are soluble in water.

Classification

Natural resins are classified according to origin into *recent* and *fossil* types.

Recent resins are obtained from living trees by making incisions and then collecting the exuding liquid in containers attached to the trunk. The depth of cut varies with the type of tree. Damar occurs in the bark and requires shallow cuts, whereas Manila occurs in the sapwood and requires deeper incisions. Resins obtained in this way are soft and are used in spirit varnishes.

Fossil resins occur as lumps in the ground and were exuded by trees which have long since rotted away. The cost of collection of fossil resins has now rendered them uneconomic for most purposes, and their use has been discontinued except for a few very specialized applications. The traditional "copal varnish" has been replaced by other types such as linseed/tung oil/ phenolic. In view of their historic interest the fossil resins are listed in Appendix A.

Composition

Resin-producing trees contain a viscous liquid or oleoresin which is exuded when the tree is cut. This liquid contains a proportion of essential oil which volatilizes on exposure to the air, and the residue hardens slowly by oxidation and polymerization.

Natural resins, in general, are very stable and not readily attacked by acids or alkalis. They are insoluble in drying oils. On heating they soften and then liquefy, but further heating causes decomposition.

PROPERTIES AND IDENTIFICATION

Appearance

They generally break with a conchoidal fracture, but a number possess a characteristic *shape* and appearance. In many cases an examination of the structure as appearing under the microscope can give useful information. For this test the resin face is polished and then etched with dilute alcoholic potash solution.

Colour

This can vary from white to very dark brown and the resins can be transparent or can exhibit varying degrees of opacity. The commercial assessment of a resin is based very largely on colour, but size of lumps and freedom from impurities are also taken into account. In one or two instances, for example, Accroides and Dragon's Blood, the colour associated with the resin was, for many years, considered a valuable asset, but these resins are now little used.

Odour

Most resins possess a characteristic odour which is intensified on slight warming.

Hardness

Natural resins differ considerably in this property. Since the majority of recent resins (with the possible exception of rosin) are used in lacquers or spirit-type varnishes, hardness is of great importance.

Melting point

With one or two exceptions, natural resins do not possess sharp melting points but soften and liquefy gradually over a range of temperature. The determination can be carried out by placing the powdered resin in a capillary tube sealed at one end and attached to the bulb of a thermometer in a heating bath. As the temperature is raised slowly, two points are noted: (i) the softening point at which the particles soften and coalesce, and (ii) the fusion point at which the material becomes liquid. The difference between these points depends to a large degree on the complexity of the resin.

The softening point of resins is often determined by the Ring and Ball softening point method set out in ASTM Method E28–51T and described by Gardner and Sward [1].

The sample is contained in a horizontal ring suspended in water which is heated slowly. The softening point is the temperature at which the sample sags through a specified distance. The apparatus is illustrated in Fig. 12.1. An 800 ml beaker A supports a brass plate B from which are suspended two

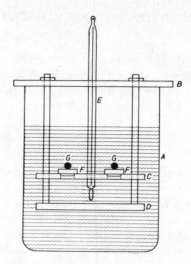

Fig. 12.1   Ring and ball apparatus

brass strips C and D. D is solid but C contains a small hole for thermometer E and two holes of larger diameter to accommodate the shouldered
brass rings F. These rings have projecting brass pins on the inside to prevent
steel balls G from dropping through. Water is used as the heating liquid.

In the determination, the brass rings are placed on a non-adherent surface such as an amalgamated brass plate, warmed, and molten resin is run
in to fill them. When completely cold they are placed in the apparatus and
the standard steel balls placed in position on the surface of the resin. The
beaker is heated slowly and at the softening point the resin will flow or sag
through the ring. The softening point is taken as the temperature at which the
resin touches plate D.

Solubility

Natural resins are all insoluble in water and show a wide diversity and
range of solubilities in organic solvents. The term *solubility* in this context
indicates mutual dispersibility rather than the formation of a true solution
of resin in solvent. Many resins absorb considerable amounts of solvent and
swell before going into solution.

Table 12.1 shows the relative solubilities of a number of natural resins.

Insoluble matter

This is important where resins are to be used in clear lacquers. It can be
determined either by allowing a resin solution of known concentration to
settle in a calibrated tube and then reading off the volume of insoluble
matter, or by filtering off the insoluble matter, drying and weighing.

Table 12.1—Solubilities of natural resins

S = soluble   P = partly soluble   I = insoluble

Resin	Alcohols	Esters	Ketones	Hydro-carbons	Mixed solvents in which resin is soluble
Rosin (colophony)	S	S	S	S	—
Mastic	S	S	S	S	—
Damar	P	S	P	S	Alcohols + esters Alcohols + hydrocarbons
Shellac	S	I	S	I	Alcohols + esters
Sandarac	S	I	S	I	Alcohols + esters
Soft Manilas	S	I	P	I	{ Alcohols + esters { Alcohols + hydrocarbons

Acid value

In natural resins appreciable quantities of the resin acids are present in the free state and the acid values are therefore generally high. It is common practice to carry out the determination using standard alcoholic potash or to add an excess of alcoholic potash and then to determine the excess by titration with standard acid.

Iodine and saponification values

Owing to the complex nature of natural resins, iodine and saponification values are of little use in identification.

Refractive index

This varies from 1·515 to 1·540 and is of little value in identification.

SPOT-TESTS FOR RESINS

Natural resins give characteristic colours in one or more of the following tests.

Storch–Lieberman test (for rosin)

The sample is warmed with an equal volume of acetic anhydride, allowed to cool, and the acetic anhydride layer separated by means of a pipette. Addition of a drop of 50 percent sulphuric acid gives a fugitive bluish-violet coloration if rosin is present.

Storch–Morawski test

This is a modification of the Storch–Lieberman test and is useful for identifying different resins. In the test, about 2 g of resin are dissolved in 10 ml of acetic anhydride by heating. The solution is cooled, placed in a white dish and treated with a drop of sulphuric acid. The results obtained with different resins are indicated in Table 12.2.

Halphen–Hicks test

Two solutions are prepared:

(i) Phenol (1 vol.) in carbon tetrachloride (2 vol.)
(ii) Bromine (1 vol.) in carbon tetrachloride (4 vol.)

A small sample of the resin is dissolved in solution (i) and a drop placed in a depression on a porcelain tile. A drop of solution (ii) is placed in an adjacent depression and the two drops are covered with a watch glass. Colorations are produced on the surface of the resin solution by the bromine evolved from solution (ii). The colours vary to some degree with different samples of the same resin, but Table 12.2 indicates the general type of result obtained.

Table 12.2

Resin	Storch–Morawski	Halphen–Hicks
Rosin	Bluish-violet (fugitive)	Green, becoming blue and finally purple
Manila	Pink	Weak green, becoming blue and finally purple
Damar	Deep red	Brown, becoming maroon
Sandarac	—	Lilac, becoming violet
Shellac	Pale green	No coloration

The natural resins in current use comprise the following:

Manila copal resins

Although the term "Manila" covers a range of resins, the material in common usage is the soft spirit-soluble type. The quality varies, some consignments containing appreciable quantities of dust and other insoluble matter. The solutions in alcohol, therefore, have to be clarified before use. These spirit varnishes are used for insulating purposes and for the manufacture of road-marking paints. The following figures are typical for the important properties:

Specific gravity 1·06–1·08   Softening point 45°C (113°F)
Acid value 145   Melting point 120°C (248°F)

Damar

A type of resin obtained from several trees of the Dipterocarpus family which is common in Malaysia, Sumatra and Borneo. The resins range from pure white to black and from transparent to opaque. They are designated by the locality of origin or port of shipment.

Damars are soft resins which melt at about 110°C (230°F) and have an acid value of 20 to 35. They are soluble in petroleum and aromatic hydrocarbons but only slightly soluble in alcohols and ketones. The solutions are generally turbid due to the presence of insoluble polymerized fractions.

They give flexible films and are therefore used in paper varnishes. They are also used extensively in cellulose nitrate lacquers to improve gloss and adhesion. For this purpose the resin must be free from any insoluble matter and a purified variety known as *dewaxed* damar is used. This is manufactured by dissolving the natural resin in toluol and adding alcohol carefully to precipitate and coagulate the polymerized material. This is filtered off and the purified damar obtained by evaporation of the solvent.

ROSIN OR COLOPHONY

Rosin, sometimes known as colophony, is the residue remaining in the stills when the oleo-resin from pine trees is distilled for the preparation of terpene solvents. The molten rosin is discharged while still hot (about 155°C, 311°F), strained, and filled into metal drums where it solidifies. The ratio of rosin to volatile solvent obtained in this way is about four to one. The product is often referred to as "gum rosin".

Wood rosin is obtained by extraction of the old stumps of pine trees which are reduced to chips and then treated with steam to separate volatile constituents. The chips are next treated with petroleum solvents to extract the rosin and the latter is obtained on evaporation of the solvent. Wood rosin is generally of inferior colour to gum rosin and contains more unsaponifiable matter.

Sources

The major part of the world's supply of rosin is obtained from the United States where the Naval Stores industry handles rosin and its derivatives. Smaller quantities are obtained from France, Portugal and Spain. A certain amount is also produced in India.

Classification

Rosin varies considerably in quality and colour, ranging from virtually colourless and transparent through shades of yellow to brown and finally almost black. The American rosin is graded by reference to a series of coloured cubes and designated by letters, e.g. WW (water-white), WG (window-glass), X (extra pale). In France the paleness of the rosin is denoted by one or more A's, up to six, the latter being the palest.

Constitution

Rosin contains about 90 percent of acids, largely in the free condition with a small amount of resenes and esters. The acids are abietic, dextro-pimaric, and laevopimaric, the structures of which are as shown overleaf.

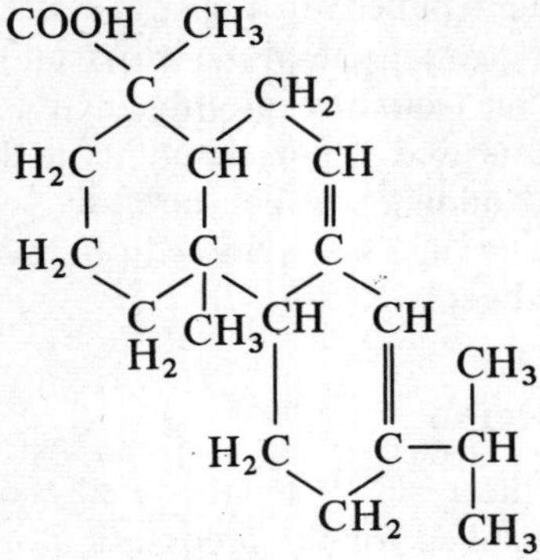

COOH CH$_3$
C
CH$_2$
H$_2$C CH CH
H$_2$C C C
C CH$_3$CH CH
H$_2$
CH$_3$
H$_2$C C—CH
CH$_2$ CH$_3$

Abietic acid

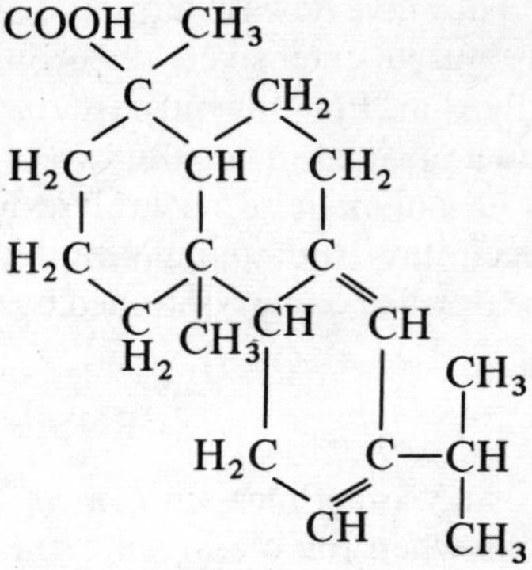

COOH CH$_3$
C
CH$_2$
H$_2$C CH CH$_2$
H$_2$C C C
C CH$_3$CH CH
H$_2$
CH$_3$
H$_2$C C—CH
CH CH$_3$

Laevopimaric acid

COOH CH$_3$
C
CH
H$_2$C CH CH
H$_2$C C C
C CH$_3$CH CH
H$_2$
H$_2$C C—CH=CH$_2$
CH$_2$ CH$_3$

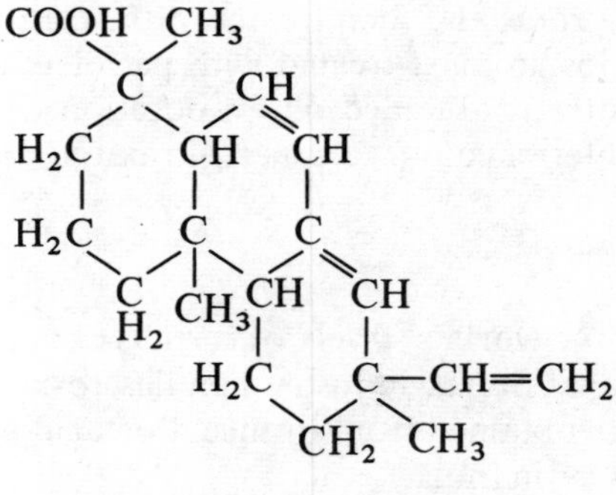

Dextropimaric acid

The proportion of abietic acid to its isomers varies according to the source of the rosin. Their unsaturated nature accounts for the fact that on exposure to the air rosin oxidizes slowly and becomes partly insoluble.

Properties

When cold, rosin is a brittle solid and small pieces can be powdered by the fingers. On warming it softens, emits a characteristic smell and melts at 110°C (230°F). The general characteristics are set out below. As a result of the high acid content, rosin forms esters (ester gums) with glycerol or pentaerythritol and also forms metal rosinates with many metals. Sodium rosinate is soluble in water, but the salts of heavier metals such as calcium and zinc are insoluble in water, though they dissolve in many organic media.

Rosin also reacts with maleic anhydride in a manner similar to unsaturated oils to produce adducts known as "maleic resins". These are used extensively in varnishes and lacquers.

The majority of organic solvents dissolve rosin when it is fresh, i.e. from the drum, but long exposure to the air results in the formation of insoluble oxidation products.

Specific gravity 1·04–1·08   Acid value 150–180.
Softening point 80°C (176°F)   Saponification value 145–195.
Melting point 110°C (230°F)   Iodine value 160–250 (usually about 230)

Uses

When made into a varnish with tung oil, rosin checks the gelling tendency of the oil; the resulting varnish is quick drying and is sometimes used for machinery paints. Considerable quantities of rosin are consumed in the manufacture of modified phenolic resins (Chapter 13).

The intense reactivity of rosin is due to the presence of a carboxyl group and two double bonds. If one or other of these is modified, considerably improved products are obtained.

Metallic salts (rosinates)

If rosin is heated with an aqueous solution of sodium hydroxide or sodium carbonate a cloudy solution of sodium rosinate (largely sodium abietate) is formed. Sodium rosinate can be used as a source of other metal rosinates. Lead, cobalt and manganese rosinates are precipitated when a solution of a salt of the metal is mixed with sodium rosinate solution. The rosinates of these metals were formerly used as driers but have now been replaced by naphthenates and octoates for two reasons: (i) the latter are more readily soluble in paint media, and (ii) the presence of rosin detracts from the durability of the paint film.

Calcium rosinate (limed or lime-hardened rosin). This is the most widely used metal rosinate and can be prepared by either fusion or solution processes. In the former the rosin is melted and heated to about 230°C (446°F), preferably in an aluminium pot to prevent discoloration. Pure lime is then added, the quantity being about 6 to 7 percent of the weight of the rosin. This is insufficient to neutralize the rosin completely, as this is undesirable, the solubility of the rosinate decreasing with high metal contents. Theoretically, about 9 percent of lime would be required for complete reaction.

The solution method consists in dissolving the rosin in white spirit and heating the solution with calcium hydroxide. Calcium rosinate is soluble in white spirit.

When calcium rosinate is used in the manufacture of wood oil varnishes, it is usual to lime the rosin in the varnish pot before addition of oil to the hot melt.

With a 6 to 7 percent calcium content the melting point of the calcium rosinate is about 125°C (275°F) and the acid value about 40.

Zinc rosinate can be prepared by heating rosin with zinc oxide, using methods similar to those for calcium rosinate. The reaction can be very vigorous and is modified by using a mixture of zinc oxide and lime.

Gloss oils are solutions of calcium or zinc rosinate in white spirit and are used in the preparation of cold-cut media where durability is of secondary importance. They are compatible with most drying oils and drying media but tend to *feed* or thicken with basic pigments.

Ester gums

These are produced by esterification of rosin with either glycerol or pentaerythritol. The glycerol ester has been known the longer and is produced by melting rosin in an aluminium or stainless steel vessel, adding 10 to 12 percent of glycerol and keeping the temperature at about 280°C (536°F) until the desired acid value is reached. This usually lies between 6 and 12. The palest types of ester gum are produced in closed vessels through which a stream of carbon dioxide is passed.

The pentaerythritol esters are produced in a similar way but at about 300°C (572°F). These esters give harder and more durable films than the glycerol types.

One of the principal uses of ester gum is in the preparation of varnishes with tung oil. These have been used extensively in gloss paints in the past, but their use today is limited to certain types of paint such as machinery finishes.

Ester gums are added to cellulose nitrate lacquers to improve gloss and adhesion. They are components of some modified phenolic resins and can be reacted with maleic anhydride.

Other rosin esters. Rosin reacts with lower alcohols such as methyl and ethyl alcohols to produce viscous liquids which are sometimes used as plasticizers. A semi-solid is produced by esterification with ethylene glycol.

Hydrogenated rosin. The degree of unsaturation of rosin is reduced by treatment with hydrogen in the presence of a catalyst. This hydrogenated rosin is less reactive and does not tend to oxidize on storage. It can be esterified with glycerol to yield exceptionally pale ester gums.

Polymerized rosin. Rosin will react with concentrated sulphuric acid to produce a polymer of higher viscosity. This will also produce spirit varnishes of higher viscosity than those produced with normal rosin. Polymerized rosin can be converted to ester gums and these yield varnishes of increased hardness and durability.

LAC (SHELLAC)

This resin occurs on the twigs and branches of certain trees and is a product of the life cycle of a particular type of insect. India produces about 95 percent of the total world consumption and the most important trees involved are the Kusum (*Schleichera trijuga*), the Palas (*Butea frondosa*) and the Ber (*Zizyphus jujuba*).

The larvae of the insect *Lacifer lacca*, which is related to the cochineal, swarm on the thin twigs of the trees and puncture the bark in order to feed on the sap. They excrete a resin which forms a protective coating over their bodies. Breathing takes place through ducts in the resinous layer, the ducts being lined with wax which is also excreted by the insect. The twigs become thickly encrusted with a dark red resin, the composition of which varies to some extent with different host trees. At the appropriate season the twigs are severed and collected. They constitute *sticklac*, the crude raw material for subsequent processing.

Seed lac. Removal of the crude resin from the twigs is carried out by a process of rolling and sifting. The crude resin is then placed in shallow tubs and well kneaded with warm water to remove the water-soluble colouring-matter and other impurities. The process is repeated and the purified resin is dried quickly in the sun to give the granular seed lac. The product varies in colour from pale yellow to brown, due to a water-insoluble colouring-matter, and is graded according to size and colour. Mechanical methods of extraction are now superseding the above traditional method.

Shellac. For many years it has been customary to add a small quantity of orpiment (arsenic sulphide) to seed lac to improve the brilliance before further processing. Rosin is also added to lower the melting point.

In the manufacture of shellac, the treated seed lac is placed in a tubular bag, about 50 mm in diameter and 6 to 9 metres (20 to 30 ft) long. This is held before a fire, when the molten lac exudes through the walls of the bag and drops on to a warm surface. The bag is slowly twisted in opposite directions from each end to assist the exudation of the resin. The plastic mass so obtained is worked out into a thin sheet about 1200 mm (4 ft) square, allowed to cool and broken up into flakes. These constitute commercial shellac, the colour of which ranges from pale yellow to brown. It is marketed in four main grades, known as Lemon (pale yellow), Fine Orange, T.N. Orange (free from orpiment and rosin), and London T.N. (brown and containing about 3 percent rosin).

Button lac differs from shellac in physical form only. The hot lac is allowed to fall on to a cold surface where it solidifies to flat discs ("buttons").

Garnet lac. The residue left in the bag after extrusion of the lac is known as "kiri" and contains the less fusible constituents and traces of dye. The kiri is removed, dried, and broken into pieces. This material is not used in the paint industry, the main outlets being a binding agent for abrasive wheels, and insulation. Mechanical methods for working up the seed lac produce good quality lacs and these are often known as "garnet lac".

Dewaxed lac. For many purposes, such as French polish for example, the presence of wax in the shellac is an advantage, but for many spirit varnishes it is undesirable. It can be removed by dissolving shellac in alcohol to give an approximately 30 percent solution, adding half the volume of white spirit, and warming to 80°C (176°F). Two layers separate, with the wax dissolved in the white spirit.

White (bleached) lac is prepared in the following manner. The lac is dissolved in sodium carbonate solution, and sodium hypochlorite solution is added slowly to decompose the colouring-matter. The solution is filtered to remove the wax and the white lac precipitated by addition of dilute sulphuric acid. It is then coagulated by heating, washed thoroughly with hot water, and moulded into hanks while still plastic; or cooled, dried and coarsely powdered. The hanks usually contain about 25 percent moisture and the powder form about 6 percent.

Bleached lac is used in the manufacture of colourless spirit varnishes, but the solubility in alcohol decreases appreciably on long exposure to the air.

Lac wax is recovered in the production of white (bleached) lac. It is a hard material and is used in some kinds of polish.

Composition of lac

The constitution of lac is very complex. It contains several substances of differing degrees of hardness and solubility and, in addition to two or more resins, contains the alcohol-insoluble wax. Two of the constituent acids have been isolated. These are aleuritic acid, present to the extent of 40 to 50 percent, and shellolic acid which is present in small quantity.

Properties and uses

Lac is a hard brittle resin which softens at about 65°C (149°F) and melts at about 80°C (176°F). The physical and chemical characteristics are set out below. It is freely soluble in alcohols and some ketones but insoluble in aliphatic and aromatic hydrocarbons, esters, and drying oils. The alcohol solution of normal refined lac is turbid, due to the presence of insoluble wax, but solutions of the dewaxed resin are clear.

The principal use of shellac in the surface coatings industry is in the manufacture of spirit lacquers and polishes. These give films which are hard, adherent, and abrasion resistant. The insolubility of lac in other media is utilized in knotting, which is an alcohol solution of lac. It is used as a sealer on new wood to prevent rosin exuding from knots into the paint film and causing discoloration. For this purpose the lac should be free from rosin.

Lac films possess good insulating properties and the resin is widely used for this purpose. It is, however, very brittle and requires a plasticizer. For this purpose methylcyclohexanol phthalate is often used. The lac films possess poor water resistance and perish rapidly on exposure to the weather. Water absorption results in a whitening of the surface which is followed by breakdown.

If alcohol solutions of lac are stored in metal containers, darkening takes place as a result of attack on the metal. This can be prevented by the use of a small quantity of ethyl aniline phosphate.

Characteristics of shellac:

Softening point 65°C (149°F) Specific gravity 1·15–1·20
Melting point 80°C (176°F) Iodine value 18–20.
Acid value 65–75. Saponification value 225–230.

Rosin adulteration is indicated by an appreciable increase in iodine value.

OTHER RESINS

Mastic is obtained as an exudation when cuts are made in the stem and branches of a Mediterranean shrub *Pistachia lentiscus* and is marketed in the form of pale yellow "tears". It is brittle and breaks with a conchoidal fracture. When placed in the mouth it softens readily and in this respect differs from Sandarac which is similar in appearance. It is completely soluble in alcohol, acetone and aromatic hydrocarbons but insoluble in white spirit. The latter property distinguishes it from Damar which is soluble.

Mastic is used in paper varnishes and for varnishing pictures. An alcohol solution is used as a fixative for pencil and crayon drawings.

Sandarac. An exudation from the tree *Callitris quadrivalvis*, a native of North Africa. Trees of similar type occur in Australia and yield a resin of practically identical character. The resin is marketed as yellowish-white tears, similar in appearance to mastic, but it powders and does not soften when placed between the teeth. It is soluble in alcohol and yields harder films than mastic. The main use of sandarac is in varnishes for charts and labels.

Reference to Chapter 12

[1] GARDNER, H. A. & SWARD, G. G., *Physical and Chemical Examination of Paints, Varnishes, Lacquers and Colors*, 12th edn, 1962. Gardner Laboratory Inc., Bethesda 14, Maryland, U.S.A.

13   Synthetic resins

The very wide range of synthetic resins available to both the paint and plastics industries is the result of systematic investigation into the formation of resinous materials, which at one time were regarded as embarrassments in chemical laboratories. Such materials showed no tendency to crystallize and had no sharp melting point. It was realized gradually that these resinous bodies were materials of very high molecular weight and that their formation was associated with the presence of certain groups in the reacting molecules. Furthermore, resin formation depended on the relative quantities of reactants.

POLYMERIZATION

Modern synthetic resins are essentially polymers, i.e. they contain large molecules which have been built up from the *normal* molecules or monomers. The number of monomer molecules which go to a given polymer molecule is not constant, so that in a resin there is a range of polymer sizes or molecular weights. These are found to be distributed round a mean value, in much the same way as the particles in a pigment.

There are two general methods for the production of polymers. These are known as *addition* and *condensation* polymerization. The method employed depends on the type of monomer.

Addition polymerization

Certain types of monomer can be reacted — usually with the aid of a catalyst — to give polymers by simple addition. The reaction usually involves addition at double bonds, and an example is afforded by the polymerization of vinyl acetate. Long-chain polymers are built up by rupture of the double bonds:

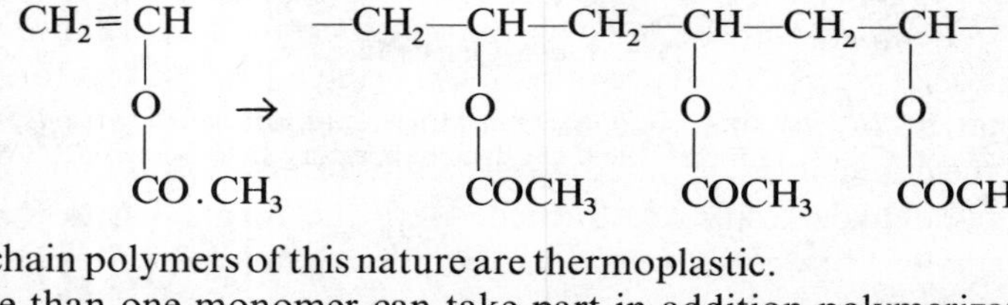

$$\begin{array}{cccc}
CH_2{=}CH & -CH_2{-}CH{-}CH_2{-}CH{-}CH_2{-}CH{-} \\
\mid & \mid \qquad\quad \mid \qquad\quad \mid \\
O \quad \rightarrow & O \qquad\quad O \qquad\quad O \\
\mid & \mid \qquad\quad \mid \qquad\quad \mid \\
CO.CH_3 & COCH_3 \quad COCH_3 \quad COCH_3
\end{array}$$

Long-chain polymers of this nature are thermoplastic.

More than one monomer can take part in addition polymerization, and vinyl acetate can be co-polymerized with, for example, butyl maleate to give a co-polymer:

$$
\begin{array}{cc}
\underset{\displaystyle\underset{\displaystyle COCH_3}{\overset{|}{\underset{|}{O}}}}{CH_2=CH} & \underset{\displaystyle\underset{\displaystyle OBu\ OBu}{\overset{|\quad|}{\underset{|\quad|}{CO\ CO}}}}{CH=CH}\ \rightarrow
\end{array}
$$

$$
-CH_2-CH\!-\!\!-\!CH-CH-CH_2-CH\!-\!\!-\!CH-CH-
$$

$$
\underset{COCH_3\ OBu\ OBu}{O\qquad CO\ CO}\qquad \underset{COCH_3\ OBu\ OBu}{O\qquad CO\ CO}
$$

The addition polymerization reaction involving the ethylene group $\diagup C=C\diagdown$ is the basic reaction in the production of vinyl, acrylic and polystyrene resins. However, the ease with which the ethylene linkage is broken to produce polymers varies with the nature of the attached groups. Ethylene and propylene are much more difficult to polymerize than vinyl acetate.

Condensation polymerization

In a very simple reaction such as that between a monobasic acid and a monohydric alcohol, water is eliminated and a simple ester is produced:

$$CH_3COOH + C_2H_5OH \rightarrow CH_3COOC_2H_5 + H_2O$$

No further reaction takes place. If, however, a dibasic acid is used with a dihydric alcohol a more complicated series of reactions can take place:

$$
\underset{CH_2COOH}{\overset{CH_2COOH}{|}} + \underset{HO.CH_2}{\overset{HO.CH_2}{|}} \rightarrow \underset{CH_2COOH}{\overset{CH_2COOCH_2CH_2OH}{|}} \qquad \textit{Half ester}
$$

This half ester can then lose water to give the cyclic ester:

$$
\begin{array}{c}
CH_2\!-\!CO\!-\!O\!-\!CH_2 \\
|\qquad\qquad\qquad| \\
CH_2\!-\!CO\!-\!O\!-\!CH_2
\end{array}
$$

or it can react with further molecules of either acid or alcohol. In this way quite large molecules can be built up by a series of esterification reactions involving loss of water. This type of reaction is the basis of alkyd formation.

Esterification is but one type of condensation polymerization reaction involving the elimination of water. Another important type is that by which phenolformaldehyde resins are formed. Here the formaldehyde reacts to form a methylene bridge between adjacent phenol molecules. In its simplest form the reaction can be represented as follows:

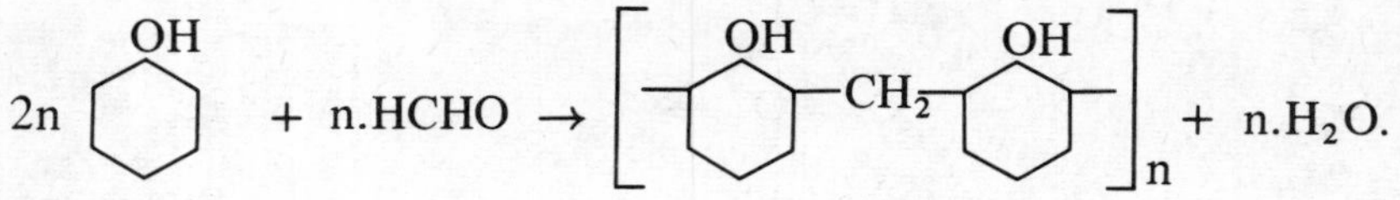

$$2n \bigcirc^{OH} + n.HCHO \rightarrow \left[\bigcirc^{OH} -CH_2 \bigcirc^{OH} \right]_n + n.H_2O.$$

Functionality

In a simple esterification such as that between acetic acid and ethyl alcohols, each reactant molecule has one reactive or functional group. The product, ethyl acetate, has no reactive or functional groups and is incapable of reacting further. The functionality is zero.

A dibasic acid such as succinic has two reactive or functional groups and so a functionality of two. Similarly, ethylene glycol has a functionality of two. The half ester produced as the first reaction stage (*above*) also has two reactive or functional groups and therefore a functionality of two.

Maleic acid contains two functional carboxyl groups and also an ethylene linkage. This linkage is capable of addition at each carbon atom, i.e., it has two reactive points, and is bifunctional. The total functionality of maleic acid is therefore four. Table 13.1 shows the functionality of a number of common reactants.

The idea of functionality was first advanced by Kienle [1] as a result of his work in alkyds, and was later developed further by Carothers [2]. The general principles relating functionality and polymer formation can be summarized as follows.

A molecule of a monomer containing two functional groups can react with a suitable second bifunctional molecule to give a polymer. Such a polymer will be linear and thermoplastic. The type of polymer, i.e. the molecular weight, will depend on the molecular ratio of the reactants. At a 1:1 ratio a high molecular weight is produced, but an excess of one will stop the reaction and lead to low molecular-weight polymers.

When a bifunctional molecule reacts with one which is trifunctional it is possible for cross-linking to occur, leading to a three-dimensional structure. Such polymers are further cross-linked on heating and yield insoluble and infusible films when stoved.

The basic reaction in alkyd preparation is the reaction between bifunctional phthalic acid and trifunctional glycerine or tetrafunctional pentaerythritol. A condensation polymerization of this type can lead rapidly to insoluble polymers and is discussed under Alkyd resins (*below*).

Table 13.1—Functional groups

Material	Formula	Functional groups	Functionality
Glycerol	CH_2OH		
	$\mid$		
	$CH.OH$	—OH (3)	3
	$\mid$		
	CH_2OH		
Pentaerythritol	$C(CH_2OH)_4$	—OH (4)	4
Phthalic acids	$C_6H_4(COOH)_2$	—COOH (2)	2
Adipic acid	$COOH(CH_2)_4COOH$	—COOH (2)	2
Sebacic acid	$COOH(CH_2)_8COOH$	—COOH (2)	2
Maleic acid	CH—$COOH$	—COOH (2)	
	$\parallel$		
	CH—$COOH$	$\rangle C=C\langle$ (2 points)	4*
Fumaric acid	CH—$COOH$	—COOH (2)	
	$\parallel$		
	$COOH$—CH	$\rangle C=C\langle$ (2 points)	4*
Formaldehyde	CH_2O or CH_2OH	—OH (2)	2
	$\qquad\qquad OH$		
	(hydrate)		
Phenol	OH	2 ortho reactive	
		1 para positions	3
Styrene			
	$CH=CH_2$	$\rangle C=C\langle$ (2 points)	2

* This means a functionality of 2 for esterification and 2 for addition reactions.

ALKYD RESINS

The alkyd resins form the largest group of synthetic resins available to the paint industry and the consumption is greater than that of any other resin. They form part of the larger group of polyesters which include all products of the esterification of polybasic acids with polyhydric alcohols.

Alkyds differ from other polyesters in containing a monobasic fatty acid in the molecule and are often referred to as *oil-modified alkyds*, though the adjective "oil-modified" is redundant. The nature and often the amount of the monobasic fatty acid determines the drying characteristics of the resin. Saturated fatty acids yield non-drying or plasticizing resins, while drying properties are conferred by unsaturated acids.

Raw materials for alkyds

The following are among the most important in common use.

Dibasic acids

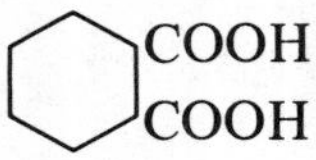

ortho-phthalic
(used as anhydride
m.p. 130°C, 266°F)

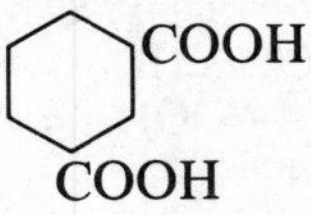

iso-phthalic
(m.p. >300°C, 572°F)

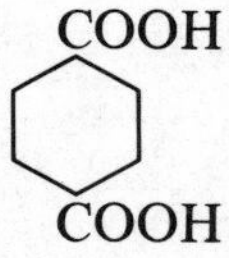

terephthalic
(sublimes on
heating)

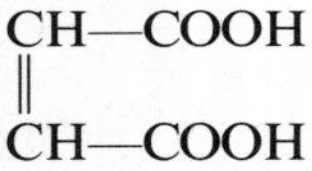

Maleic
(m.p. 130°C,
266°F)
(anhydride
m.p. 56°C,
132·8°F)

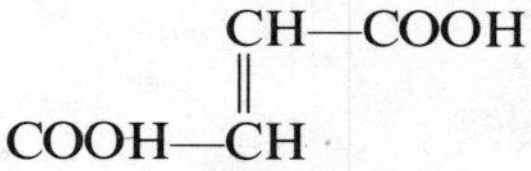

Fumaric
(sublimes at
200°C, 392°F)

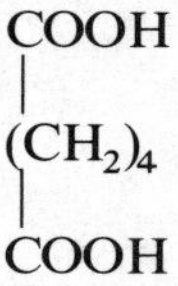

Adipic
(m.p. 151°C,
303·8°F)

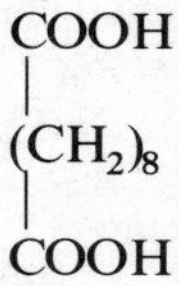

Sebacic
(m.p. 133°C,
271·4°F)

Orthophthalic acid, used in the form of the anhydride, is the most extensively used of these acids.

Polyhydric alcohols (polyols). The two most widely used are glycerol and pentaerythritol, but small amounts of sorbitol and trimethylolpropane are sometimes employed.

Glycerol. The general properties of glycerol have been discussed in Chapter 10. It contains two primary and one secondary alcohol group and these react at the same rate with fatty acids to form glycerides. With dibasic acids the primary groups react more readily than the secondary; and it is likely that the former react with phthalic anhydride to give the first stage syrup, and the secondary alcohol groups are available for reaction with fatty acids.

Glycerol can be used in alkyds of all fatty acid contents or oil lengths.

Pentaerythritol, $C(CH_2OH)_4$, is a solid with melting point 253°C (487°F), and is manufactured by reaction between acetaldehyde and formaldehyde in the presence of water and alkali. It contains four primary alcohol groups which are all equally reactive and forms more complex resins than does glycerol. It can be used for medium/long and long oil alkyds only, as the short and medium/short oil types gel rapidly. Compared with the glycerol types, long oil alkyds produced from pentaerythritol dry harder to give more durable films with better moisture resistance. They have a greater tendency to form crystalline bloom (see Chapter 6, Vol. 2) in urban atmospheres.

Fatty acids. The general properties of the fatty acids have been discussed in Chapter 10.

It has been mentioned earlier that the drying characteristics of an alkyd are determined by the quantity and nature of the fatty acids, and they can be divided into two general groups, drying and non-drying.

The acids employed consist of mixtures derived from the parent oil by hydrolysis of the glycerides. In this way the acids can be separated and, if necessary, purified. An alternative method in which the acids are not isolated consists in heating together the oil and glycerol to produce monoglycerides as in the process of this name (*below*).

Alkyd formation

When two moles of glycerol and three moles of phthalic anhydride are heated together, a syrup is first formed which, on heating further, is converted to a gel. Continued heating causes the formation of a resin which ultimately becomes an insoluble and infusible mass. It is thought that a linear polymer is first formed, and this gradually cross-links to give a very complex structure. This is the type of resin usually formed from a bifunctional acid and a trifunctional polyol and is of no value as a surface-coating resin.

The polymerization to the very large polymers can be checked by the introduction of a monofunctional component into the system. This can be as an alcohol such as cyclohexanol or *n*-butanol, or as a monobasic fatty acid. In practice the latter is used and the fatty acid can be reacted with the first stage or syrup formed from phthalic anhydride and glycerol. On further heating the resin is much less reactive and does not reach the infusible state. This is the basis of the fatty acid method of alkyd manufacture.

Fatty acid process

There are two variants of this process:

(i) The phthalic anhydride and glycerol are heated to 180°C (356°F) to the "first stage" syrup, and molten fatty acids are added to esterify the free hydroxyl groups. Heating is continued at 180° to 220°C (356° to 428°F) until the desired acid value and solubility characteristics are reached. Foaming can be troublesome in this method.

(ii) The three raw materials, phthalic anhydride, glycerol and fatty acid, are placed in the reaction vessel together with a small quantity of xylol. The vessel is fitted with a condenser to which is attached a water separator of the Dean and Stark type. On heating, the water produced is carried off with the xylol and is separated. The condensing xylol serves to flush the sublimed phthalic anhydride back into the reaction vessel. The amount of water collected is an indication of the progress of esterification, but samples of the resin are removed from time to time for acid value and viscosity checks.

The reaction vessels are generally of stainless steel fitted with stirrer, charge hole, condensing system, and pipes for passing inert gas over the charge. The latter serves to reduce discoloration. Heating can be by immersion heaters or by passing hot oil or other liquid through a jacket surrounding the vessel.

The fatty acid process is being superseded by the alcoholysis process (*below*).

Alcoholysis or monoglyceride process

The formation of an alkyd resin involves esterification of glycerol or pentaerythritol with phthalic anhydride and a fatty acid. In glycerol-type alkyds (but not those from pentaerythritol — see below) the necessity for isolating the fatty acids can be avoided by the use of monoglycerides which are then further esterified with phthalic anhydride. The monoglycerides are formed by heating the drying oil with the necessary amount of glycerol to 250° to 280°C (482° to 536°F), when alcoholysis takes place according to the scheme:

$$
\begin{array}{ccccc}
CH_2OCOR & & CH_2OH & & CH_2OH \\
| & & | & & | \\
CHOCOR & +\ 2 & CHOH & \rightarrow\ 3 & CHOH \\
| & & | & & | \\
CH_2OCOR & & CH_2OH & & CH_2OCOR
\end{array}
$$

The phthalic anhydride is added and the reaction completed at 180° to 250°C (356° to 482°F). The plant employed is similar to that used for the fatty acid process and the reaction is assisted by the use of xylol to remove the water.

This process is cheaper than that using the fatty acids, but it is not as versatile. A greater variety in the nature and purity of acids is possible with the fatty acid method.

Alkyds required to contain pentaerythritol as the sole polyol cannot be

produced by the alcoholysis or monoglyceride process since the product
will contain the glycerol introduced with the monoglyceride.

Dehydrated castor oil alkyds

These differ from other alkyds in their rather unique method of manufacture. The raw materials — raw castor oil, phthalic anhydride and glycerol
— are heated together, slowly at first and then rapidly to 260° to 270°C
(500° to 518°F). Under these conditions the raw castor oil is dehydrated
by the phthalic anhydride, leading to the formation of dehydrated castor
oil resins. It is of interest that during the dehydration reaction the viscosity
of the mix falls, but rises again when polymerization of the resin commences.

Drying alkyds

These contain unsaturated fatty acids which control, to a large degree, the
drying and performance characteristics of the resin. The amount of fatty
acid present in the glycerol type can vary from 25 to 80 percent and is generally expressed as a percentage, although the varnish convention of long-,
medium- and short-oil is sometimes used. The figures covered by these
terms are not precise, but they refer roughly to fatty acid contents of 60
percent upwards, 45 to 60 percent, and 25 to 45 percent respectively.

Pentaerythritol is used only in alkyds with a fatty acid content of about
60 percent and upwards. It tends to gel rapidly when used with lower proportions of fatty acids.

Drying alkyds with fatty acid contents of 60 percent and above are
soluble in aliphatic hydrocarbons such as white spirit, and are compatible
with drying oils and varnishes. In the range of 45 to 60 percent fatty acid
a proportion of stronger aromatic solvent is often necessary, but below 45
percent fatty acids the resins are soluble only in strong solvents such as
xylol and naphtha. They are not compatible with drying oils or varnishes.

Driers

The drying times decrease progressively with decrease in fatty acid content, the actual figures depending on the type of acid used. The long and
medium types are catalysed satisfactorily by mixtures of lead and cobalt in
the form of naphthenates and octoates. The use of lead, however, is being
discontinued and the metal has been eliminated from decorative paints;
these dry satisfactorily with zirconium/cobalt mixtures. In decorative finishes
a proportion of calcium (usually as octoate) is included to check the tendency
to "crystalline bloom" formation on exterior films. The short types of alkyd
can be catalysed by cobalt alone, and this type is frequently cured by stoving
without driers.

Types of fatty acid employed

Linseed fatty acids produce alkyds with good drying properties and durability but which yellow to a marked degree on exposure. This yellowing appears to be associated with the linolenic acid present. Linseed alkyds based on pentaerythritol dry harder and are more durable than those based on glycerol.

Soya bean fatty acids are widely used and, by reason of their very low linolenic acid content, the tendency to yellow on exposure is much less than with linseed. Alkyds made with these fatty acids and pentaerythritol as the polyol are in general use in the pale shades of decorative enamels.

Safflower, *Sunflower* and *Tobacco* seeds yield fatty acids free from linolenic acid and containing a high proportion of linoleic acid (like soya bean oil, they are described as "linoleic rich"), and are used as alternatives to soya acids.

Dehydrated castor oil fatty acids (a mixture of 9:11-octadecadienoic and 9:12-octadecadienoic acids) yield non-yellowing alkyds. These are made in medium- and short-oil lengths and are used in air-drying and stoving (with amino resins) finishes respectively. These fatty acids are also used in some epoxy esters.

Tall-oil fatty acids (essentially linoleic) are derived from the wood-pulp industry. They can now be produced virtually free from rosin acids and yield non-yellowing alkyds exhibiting excellent gloss.

Tung-oil fatty acids are not used alone but often mixed with linseed in alkyds which can be further modified with phenolic resin. Such alkyds dry faster than the linseed type and the film has much greater water resistance. The durability is very good but the film yellows on exposure.

Applications of drying alkyds

Alkyd finishes, in general, show good gloss and gloss retention but are sensitive to moisture both during drying and afterwards. Exterior painting with alkyds during conditions of condensation, which often occur in early summer and autumn, can lead to loss of gloss on account of microrivelling.

When immersed in water for prolonged periods, alkyd films absorb appreciable quantities of water, and the resulting swelling frequently results in blistering. They are therefore not used under conditions of water immersion.

The long and medium types are used in air-drying finishes and under-coats for decorative and maintenance work and in marine topside and weatherwork paints. The medium types are used also in paints for machinery and implements.

Long-oil alkyds are versatile materials, and some are compatible with other types of resin to give films with special properties. Blends with chlorinated rubber, for example, show greater adhesion to clean metal than does chlorinated rubber alone, and are sometimes used in primers for steelwork under chlorinated rubber systems.

Mixtures of some long-oil alkyds and certain vinyl co-polymers show rapid through-drying and are used in undercoats when short overcoating times are required. Care must be taken in the choice of resins and solvents since these mixtures can present stability problems.

Short-oil types find use in quick-drying primers and sometimes in "half-hour" enamels. They are used also in stoving coatings often with amino resins.

Non-drying alkyds

These are produced either by the use of saturated fatty acids in the normal process, or by the polymerization of a bifunctional acid such as adipic with a bifunctional polyol such as ethylene glycol. The former types are the more common, and resins based on castor oil are very widely used. These are made by heating raw castor oil, glycerol and phthalic anhydride at a relatively low temperature (190° to 200°C, 374° to 392°F).

Non-drying alkyds are made from other oils by the fatty acid or alcoholysis process and are usually of short oil length.

They are used as plasticizers for cellulose nitrate lacquers and as film-formers in conjunction with amino resins. In the latter case the films are stoved to effect the necessary cross-linking between the resins.

The non-drying alkyds used in high quality white stoving enamels are frequently based on lauric acid or coconut oil fatty acids.

Vinyl toluenated (VT) alkyds

Vinyl toluene, $CH_3C_6H_4CH=CH_2$ (as well as styrene, $C_6H_5CH=CH_2$) will react with alkyds containing unsaturated linkages when heated in xylol solution in presence of a catalyst (e.g. a peroxide). Addition of vinyl toluene takes place at the double bonds in the fatty acid chain and has the effect of reducing the oil length. The products dry more rapidly and give harder and more chemically resistant films. They can take some time to develop their solvent resistance, and recoating can sometimes be a problem.

Vinyl toluenated alkyds are used for rapid air-dry and stoving finishes in industrial work. The longer oil types (with the lower vinyl toluene content) are less affected by adverse drying conditions than are the straight alkyds, but their durability is not as good.

Silicone alkyds

These are produced by co-polymerizing alkyds with certain silicone inter-mediates, e.g. low molecular weight siloxanes, so that the product contains up to 30 percent silicone. They are characterized by good drying properties, high gloss, and outstanding weather resistance. Some "long life" coating systems are based on these resins.

Urethane alkyds

These are discussed on page 228.

Thixotropic alkyds

Certain modified polyamide resins will combine with conventional alkyds to produce resins with pronounced thixotropic properties.

It is possible, by the use of suitable types and quantities of reactants, to produce resins which can be pigmented to give finishes ranging from full gloss to flat. These are the so-called **jelly** or **dripless** paints which form firm gels at rest but lose their structure and exhibit flow when stirred or applied by brush.

Water-soluble alkyds

One of the most interesting developments of recent years has been the production of alkyd (and other) resins capable of being solubilized in water, but which can be stoved to a water-resistant film. Water-solubility is achieved by the introduction of a number of free carboxyl and hydroxyl groups which are well distributed in the polymer. These groups are neutralized by organic bases, the type of which depends on the type of polymer, but, in general, the tertiary bases are preferred. The products are soluble in water and can be pigmented and applied by spraying, dipping, or electrodeposition. The film is cured by stoving, when the organic base is volatilized and cross-linking occurs.

UNSATURATED POLYESTER RESINS

Although alkyd resins are considered to be part of the polyester group, the resins considered here differ from alkyds in several respects: (i) they contain no fatty acid, (ii) dihydric alcohols, e.g. glycols, are used in place of glycerol and pentaerythritol, and (iii) they contain unsaturated bifunctional acids.

Formation

The product obtained from the condensation polymerization of a glycol such as propylene glycol, with a mixture of saturated and unsaturated dibasic acids such as phthalic and maleic (preferably as anhydrides), is a linear polymer containing unsaturation in the chains. This is dissolved in a suitable monomer, generally styrene, to produce the final resin. The film is produced by co-polymerization of the linear polymer and monomer by means of a free radical mechanism. The free radicals can be generated by heat, or more usually by addition of a peroxide. In order that curing can take place at room temperature the decomposition of peroxides into free radicals is catalysed by certain metal ions, usually cobalt. **Caution:** the solutions of peroxide and cobalt compound *must not be mixed* under any circumstances but should be added separately to the resin mix and well stirred in. A simple mixture of the two additives is highly explosive.

Films produced by curing in the air can show a residual surface tack for some hours. This is the result of inhibition of curing by the oxygen of the air and is difficult to overcome. One method consists in dissolving wax in the paint, so that the wax forms a skin on the surface of the film. When the latter is dry the wax is removed by polishing. The amount of wax used is of the order 0·01 to 1·0 percent. Wax-free types which are not inhibited by oxygen to the same degree incorporate an allyl group, such as the allyl ether of trimethylolpropane or glycerol, as part of the polyol ingredient.

Properties and uses

Polyesters give hard, mar-resistant films with good chemical and solvent resistance. The films are characterized by depth of gloss. These resins are used in considerable quantities for wood finishing.

The fact that polyesters cure by a free radical mechanism renders them suited to irradiation curing. This form of cure, in which no heat is produced, is particularly suited to films on wood or board.

Electron-beam curing is used on flat surfaces and it effects cure in less than one second. The more widespread use of electron-beam curing has been hindered by difficulties in the design of plant.

Ultraviolet irradiation curing is used on an increasing scale, particularly for clear coatings on wood or board. Difficulties have been experienced with pigmented coatings. These are discussed under "Acrylic resins" (page 232).

AMINO RESINS

These are derived by reaction between formaldehyde and either urea or melamine to give urea–formaldehyde or melamine–formaldehyde resins respectively.

$$
\begin{array}{cc}
NH_2 & \\
| & \\
CO & \\
| & \\
NH_2 &
\end{array}
$$

Urea
(m.p. 133°C, 271·4°F)

Melamine
(m.p. 350°C, 662°F)

Urea–formaldehyde resins

When urea is heated with formaldehyde solution (formalin) under slightly alkaline conditions, addition takes place to form methylol urea. Either mono- or dimethylol urea is formed according to the ratio of urea to form-aldehyde:

$$
\begin{array}{c}
NH_2 \\
| \\
CO \\
| \\
NH_2
\end{array}
+ H.CHO \rightarrow
\begin{array}{c}
NH.CH_2OH \\
| \\
CO \\
| \\
NH_2
\end{array}
\qquad \textbf{Monomethylol urea}
$$

$$
\begin{array}{c}
NH_2 \\
| \\
CO \\
| \\
NH_2
\end{array}
+ 2H.CHO \rightarrow
\begin{array}{c}
NH.CH_2OH \\
| \\
CO \\
| \\
NH.CH_2OH
\end{array}
\qquad \textbf{Dimethylol urea}
$$

A water-soluble syrup is first formed, and on continued heating to 200° to 240°C (392° to 464°F) hardens rapidly to a colourless hard infusible resin. This resin is valueless as a coating material, but if the condensation is carried out in the presence of a high-boiling primary alcohol such as *n*-butanol, the hydroxyl groups in the methylol ureas are etherified:

$$
\begin{array}{c}
NH.CH_2OH \\
| \\
CO \\
| \\
NH_2
\end{array}
+ \quad HO.C_4H_9 \quad \rightarrow
\begin{array}{c}
NH.CH_2OC_4H_9 \\
| \\
CO \\
| \\
NH_2
\end{array}
+ H_2O
$$

The butylated urea resin produced by polymerization of this ether possesses entirely different solubility properties from the original hard resin. It is soluble in aromatic and some aliphatic hydrocarbons but insoluble in water. It is miscible with some alkyds and some varnishes. The solubility and compatibility with other resins increases with increasing chain length of the alcohol. When the films are cured by stoving, much of the alcohol breaks off and is lost, but some is retained as an integral part of the cured film.

Manufacture of urea resins

Urea and formalin are charged into a stainless steel vessel fitted with stirrer, condensing system, and water collector. The pH of the mixture is adjusted to 8. It is warmed for a short time and then an excess of butanol is introduced. The butanol and water distil into the collector and separate into two layers. The upper butanol layer runs back into the reaction vessel and the water can be tapped off. Such a process is known as azeotropic distillation. The pH of the resin is adjusted to 5 by addition of phosphoric acid, and etherification takes place. The final resin is obtained as a colourless solution in *n*-butanol.

Properties and uses

Alkylated urea resins are water white and are compatible with a number of alkyds, both drying and non-drying. The combination of urea resin with non-drying alkyd is used very widely for stoving finishes of almost all types. It appears that combination between the two resins takes place during stoving, probably by etherification of the hydroxyl groups of the alkyd with the butoxy or methylol groups of the urea.

For general purposes a urea:alkyd (solids) ratio of about 1:1 is used and stoved at 120°C (248°F) for 20 to 30 minutes.

Curing of the films at lower or ambient temperatures can be effected by addition of small quantities of mineral or phosphoric acid.

Melamine–formaldehyde resins

A molecule of melamine can react with up to six molecules of formalde-hyde to give a series of methylol melamines:

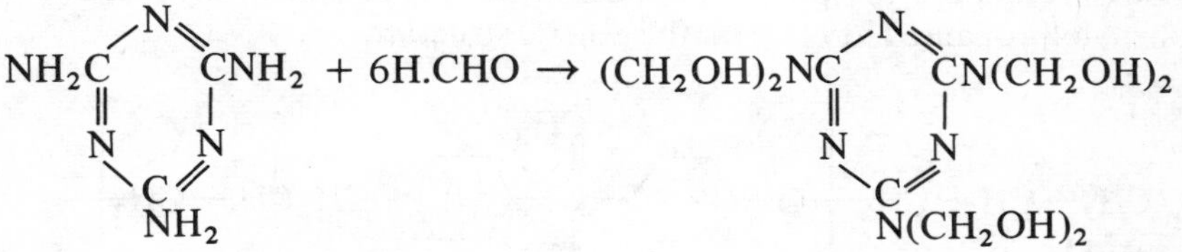

These methylol melamines are very reactive and the mono-, di-, and tri-compounds tend to react, forming methylene bridges between melamine units. They form ethers with n-butanol, and this reaction is generally employed to limit the reactivity and to confer solubility in solvents and other media.

Manufacture of melamine resins

The process employed is similar to that used for urea resins, but conditions are modified in view of the greater reactivity of the melamine.

Properties and uses

Melamine resins are supplied in either *n*-butanol or *n*-butanol/xylol solutions and have limited compatibility with white spirit. They are miscible with drying and non-drying alkyds as well as with some epoxy resins and thermosetting acrylics.

Combinations of melamine resins and non-drying alkyds are widely used in white stoving finishes for domestic equipment and in car finishes. Melamines will tolerate larger quantities of alkyd than will ureas and the films show greater gloss retention, and heat and chemical resistance. The weather resistance and durability are excellent and they are used with thermosetting acrylics in car finishes.

Melamine/alkyd finishes can be cured at slightly lower temperatures than ureas. Like the latter they can be cured at ambient temperatures by addition of small amounts of mineral or phosphoric acid.

The mechanism of cure of alkyd/melamine systems by acid has attracted much attention. A summary and critical examination of the suggested mechanisms has been published by Holmberg [3]. The same author points out [4] that under acidic curing conditions, drying-oil alkyds give harder films than the non-drying types. He explains this by a Diels–Alder cross-linking mechanism.

Water-soluble melamines with soluble alkyds are used in the electrodeposition of gloss finishes.

EPOXY RESINS

Epoxy resins are the products of condensation of epichlorhydrin and diphenylolpropane. They possess the general structure

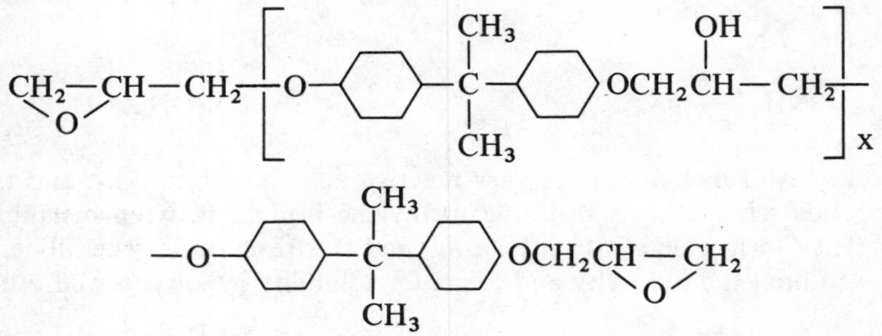

where x indicates the repeating unit. They are formed by condensing the reactants in the presence of alkali at about 100°C (212°C). A range of resins is produced, differing in molecular weight, and they are marketed under the trade names "Epikote" (Shell Chemical Company) and "Araldite" (CIBA/Geigy Ltd).

Epoxy resins contain hydroxyl and epoxy groups and the epoxy content is important in the evaluation of these resins. This may be quoted either as *epoxy equivalent*, i.e. the number of grams of resin containing one gram-equivalent of epoxy group, or as *epoxy value*, i.e. the number of epoxy groups in 100 grams of resin. The hydroxyl groups can be esterified with carboxylic acids to give epoxy esters (*below*).

Epoxy resins are thermoplastic, soluble in esters and ketones and in mixtures of these with aromatic hydrocarbons and some alcohols. They can be cross-linked to give stoving or cold-curing coatings by the use of amino or phenolic resins, by amines, or by polyamide resins. The resulting films are very resistant to chemical attack and possess a high degree of adhesion to metal and other surfaces.

Curing by amino or phenolic resins

These materials cross-link with epoxy resins on stoving to give films of excellent adhesion, chemical resistance and flexibility. In each case the cross-linking occurs through the epoxy groups. Films obtained from butylated urea resins are paler than, and possess slightly inferior resistance to, the phenolic types, but are more widely used on account of their better colour. They are used in white finishes for articles such as washing machines. The phenolic types are used where colour is of less importance, as in drum linings.

Curing by amines

Primary amines will react with epoxy resins at ambient temperatures to give long-chain polymers according to the general scheme

$$R'NH_2 + CH_2 \underset{O}{\overset{}{\diagdown\!\!\diagup}} CH.CH_2 R'' \rightarrow R'NH.CH_2.\underset{\underset{OH}{|}}{CH}\!-\!CH_2 R''$$

$$R''CH_2 CH \underset{O}{\overset{}{\diagdown\!\!\diagup}} CH_2 + R'NH.CH_2.\underset{\underset{OH}{|}}{CH}.CH_2 R''$$

$$\downarrow$$

$$R''CH_2 CH.CH_2.\underset{\underset{OH}{|}}{N}\!-\!CH_2.\underset{\underset{OH}{|}}{CH}.CH_2 R''$$
$$\underset{OH}{|} \qquad \underset{R'}{|} \qquad \underset{OH}{|}$$

The use of primary, di-, or triamines such as ethylene diamine or di-ethylene triamine leads to cross-linking and the formation of extremely resistant films. When the two components are mixed, set and cure of the film takes place at normal temperatures but can be accelerated further by the use of heat. Generally the amount of amine used is about 6 percent on the weight of the resin, and the two are mixed shortly before use. These systems are therefore supplied as two components and the mixture has a limited "pot-life". This depends on the type of solvent used, but it is unwise to mix more than is required for one day's working. The pot-life is extended slightly when the resins are pigmented.

Relatively thick films can be produced by this method of cure and they show high resistance to chemicals and solvents.

The amines used in these reactions are generally toxic and require careful handling. For this reason amine "adducts" were introduced. In these materials the amine groups are rendered innocuous so that the adduct can be handled safely, but the groups are available for the curing reaction when mixed with the epoxy resin. A typical adduct is that comprising diethylene triamine and bisphenol A diglycidyl ether.

Solvent-free systems. By the use of liquid (low molecular weight) epoxy resin and a liquid (amine) hardener it is possible to apply very thick coats and to avoid the faults which can arise from solvent retention in thick films. Drawbacks are high application viscosity and a very short pot life. This type of system is used in underwater application. It is also used in "potting" electrical components.

High solids systems incorporate a small quantity of solvent. This renders application a little easier and tends to lengthen the pot life.

Curing by polyamides

These resins (page 225) contain the reactive group $-CO.NH-$ and react with epoxy resins in a similar way to amines, but the reaction is slower and more amenable to control. The films produced are identical in many properties, but whereas polyamides yield the more flexible and water-resistant films, amines and amine adducts give films with higher chemical resistance. These differences are, on the whole, relatively small, and the lower toxicity of polyamides together with the more controllable reaction has resulted in these materials being more widely used.

Curing by isocyanates

The higher molecular-weight epoxy resins are used in order to furnish the maximum number of hydroxyl groups for reaction with the isocyanate. The resulting film shows very high chemical resistance, adhesion, and general mechanical properties.

Epoxy/coal-tar systems

The epoxy/polyamide and epoxy/amine or adduct systems can be blended with special grades of coal-tar pitch to give products which are considerably cheaper than the resins and which can be applied in thick coatings to steel and other metals to protect them from corrosion. Like the resins themselves, they are two-pack systems and are widely used in the protection of ships' bottoms and other marine structures. A drawback to this type of coating is the fact that the pitch "bleeds" into most superimposed paints.

Other applications of epoxy resins

In addition to the applications mentioned above, they are often combined with thermosetting acrylic resins (p. 233) and are also used, in admixture with sand, in flooring compositions. Large quantities of epoxy resins are used in powder coatings.

Economics

Epoxy resins are expensive and are used in circumstances where the high cost is justified by the high degree of protection obtained.

Epoxy esters

The hydroxyl groups in epoxy resins can be esterified with fatty acids derived from linseed, dehydrated castor, soya, or coconut oils to produce esters, the properties of which depend on the type of acid used. Long-, medium-, and short-oil types can be prepared.

To some extent they resemble alkyds but possess greater chemical resistance and flexibility. Long-oil and medium-oil types based on drying-oil fatty acids are used for air-drying brushing finishes, while the short-oil types are used in spraying finishes which will air-dry rapidly; they can also be stoved. The latter are compatible with amino resins and these combinations are frequently used in industrial work.

Films of epoxy esters show good adhesion, flexibility and chemical resistance. They are used under conditions of moderate chemical attack but are not as resistant as the cross-linked epoxy coatings.

POLYAMIDE RESINS

These resins are formed by condensation of polyamines with dimer (or other polybasic) acids, the nature of the product being determined by the functionality rules enunciated by Carothers.

Dimer acids are long-chain aliphatic dicarboxylic acids containing two or more alkyl side chains. They are obtained by polymerization of refined unsaturated C_{18} fatty acids, and the long chains present confer a high degree of flexibility.

Apart from their use in polyamide resins they are used as part-replacements for dibasic or fatty acids in alkyds and epoxy esters.

Polyamide resins have two important applications:

(a) They are used as curing agents for epoxy resins (p. 224). For this purpose they are more convenient to handle than the amines, but there are slight differences in the properties of the products.
(b) When reacted with alkyd resins they produce the "thixotropic" alkyds which are used in "non-drip" decorative paints. The polyamide is added during the final stage of alkyd production, and the structure is thought to result from hydrogen bonding between carboxyl and amine groups.

POLYURETHANE RESINS

The basis of formation of these polymers is reaction of an isocyanate group (–NCO) with compounds containing an active hydrogen atom. The latter attaches itself to the nitrogen atom of the –NCO group. A number of materials contain active hydrogen atoms, and the reactions of an isocyanate group with some of these can be represented as follows:

(1) *Reaction with hydroxyl group*

$$-N=C=O \ + \ -OH \ \rightarrow \ \overset{\displaystyle H}{\underset{\displaystyle |}{-N}}\overset{\displaystyle O}{\underset{\displaystyle \|}{-C}}-$$

urethane "unit"

This reaction is used in the formation of surface films. A di-isocyanate is reacted with hydroxyl-containing materials such as alkyds or polyesters.

(2) *Reaction with carboxyl group*

$$-N=C=O \ + \ HOOC- \ \rightarrow \ -NH-CO- \ + \ CO_2$$

(3) *Reaction with amine*

$$-N=C=O \ + \ H_2N- \ \rightarrow \ -NH.CO.NH-$$

(substituted urea)

(4) *Reaction with water*

$$-N=C=O \ + \ H_2O \ \rightarrow \ -NH_2 \ + \ CO_2$$

This reaction is used in foam production. The amine produced reacts immediately with further isocyanate to give a substituted urea, as in (3). The substituted urea can react further with free $-NCO$ groups to give a highly cross-linked structure.

Film formation by moisture-cured polyurethanes results from a train of reactions of this nature.

In reaction (1), the hydroxyl groups may be phenolic or aliphatic, and the latter may be primary, secondary or tertiary. Saturated or unsaturated compounds may be involved.

The isocyanates may be long- or short-chain, aliphatic, or aromatic. Similar comments apply to reactions (2) and (3). It is therefore apparent that a very great number of technical combinations are theoretically possible.

MATERIALS FOR POLYURETHANES

Polyurethanes are two-pack materials, i.e. the material containing the active hydrogen (polyester, alkyd or epoxy) and the isocyanate are packed in separate containers and mixed together immediately before use. The chemical reaction commences at once and consequently such mixed materials have a limited "pot life".

Isocyanates

Toluene di-isocyanate (TDI) is widely used in the form of polyisocyanate produced by treatment of the monomer. This, and other polyisocyanates, has a low vapour pressure and usually contains less than 0·7 percent of the highly toxic monomer (TLV 0·14 mg/m^3). Inhalation of the latter can result in serious respiratory effects and can lead to impairment of lung function [5]. The dangers associated with the use of polyisocyanates prompted the Paintmakers' Association to issue a document entitled "The Hazards arising from the Manufacture of Polyurethane Paints and Varnishes". The toxic effects are confined to the isocyanate raw materials. The resulting polyurethanes are not hazardous (apart from solvent), provided no unreacted isocyanate is present. Films produced from aromatic polyisocyanates yellow on exposure; the aliphatic compounds do not show this defect.

Resins

The resins most widely used for reaction with di-isocyanates are saturated polyesters, the general requirements being a high hydroxyl value and an acid value as low as possible to avoid liberation of carbon dioxide.

The nature of the film produced will depend on the chemical structures of both polyester and isocyanate. Soft films are formed from polyesters based on aliphatic acids, but harder and sometimes brittle films result from the use of alkyds based on phthalic and terephthalic acids. The type of polyol used in the resin also affects the film properties. Glycerol and pentaerythritol give hard films by cross-linking, but diols such as 1,4-butanediol or ethylene glycol give much softer films.

In view of the very wide range of properties, especially flexibility and hardness, required for different types of application and surface, manufacturers find it convenient to produce a limited range of polyesters of widely differing characteristics. The desired properties are then obtained by mixing the polyesters in the appropriate ratios. These polyester resins are marketed under trade names, e.g. Daltolacs (ICI) and Desmophen (Bayer).

Solvents

The polyesters used are soluble in esters and ketones, and a balanced mixture of low-, medium-, and high-boiling types is generally used. In some cases toluene or xylene is used as a diluent to reduce costs. Alcohols must be avoided. The di-isocyanates are generally dissolved in esters. All solvents should be dry to avoid reaction between water and the isocyanate group, leading to formation of carbon dioxide.

Film formation

Two-pack type

This comprises the largest class and is used for both clear and pigmented coatings. The ratio of isocyanate to polyester is important in that low ratios of isocyanates give softer films.

Calculations of reacting ratios can be difficult with polyesters which may contain more than one polyol. The di-isocyanates are fairly pure and their equivalent weight is known. The polyester is calculated on the basis of an experimental value known as *isocyanate equivalent*. This is the weight of polyester (in grams) which reacts with one gram-equivalent of isocyanate. The blend of polyesters is chosen to give the desired properties, and the isocyanate required is calculated from their isocyanate equivalents.

Polyester and isocyanate are made up in separate containers and mixed in the appropriate ratio shortly before use.

In pigmented finishes the pigment is dispersed in the polyester, and most pigments can be used safely. However, basic pigments such as red lead, zinc oxide and zinc chrome as well as Prussian Blue cause premature gelling of the system.

Additions of cellulose acetobutyrate or ethyl cellulose are often made to improve the flow properties and to avoid *cissing*.

Properties and uses

These polyurethane films can be produced in a very wide range of flexibility and hardness and also show good adhesion to a number of surfaces. They can therefore be used on rigid articles where hardness is required or on soft rubber objects where maximum flexibility is essential. The films are very resistant to extreme conditions and withstand chemical attack, heat, moisture, and many solvents. They are weather resistant but some tend to yellow.

Moisture-cured type

The polyether triols (prepared by reaction of glycerol with polypropylene glycol) contain secondary hydroxyl groups and react more slowly with isocyanates than do the normal polyesters. If these are mixed with an excess of di-isocyanate a reaction occurs giving virtually a solution of a prepolymer. Films of this material when exposed to the atmosphere react with the moisture present and cross-linking takes place.

These materials are generally used as clear coats since the moisture present in the pigments is sufficient to start the cross-linking reaction. They cure to very hard, adherent, gloss films with high abrasion resistance. Uses include floor varnishes and marine applications where traces of moisture on the surface are not detrimental.

One-pack urethane alkyds

These are prepared by forming a mixture of mono- and di-glycerides from a drying oil and reacting the free hydroxyl groups with a di-isocyanate. The product is soluble in white spirit and cures by the normal drying process, usually with the addition of cobalt drier. One-pack urethane alkyds dry to hard and tough films, but the overall resistance is not as high as that of the two-pack type of polyurethane. The film hardness and resistance to solvents and chemicals is superior to normal alkyds, and they can be used to upgrade these materials.

Other advantages of the urethane alkyds are solubility in white spirit and freedom from pigmentation difficulties.

VINYL RESINS

Vinyl resins are straight-chain thermoplastic polymers produced by addition polymerization of compounds containing the vinyl group $CH_2=CH-$. The polymers can be made from straight monomers to give homopolymers or from two or more monomers to give co-polymers. In this way a very large number of co-polymers is possible, for example by polymerizing vinyl acetate with other unsaturated monomers. The polystyrene and acrylic resins are also members of this class.

Methods of polymerization

Two methods are employed: (a) *solution polymerization* in which the monomer is dissolved in a suitable solvent, and (b) *emulsion polymerization* in which the monomer is emulsified in a non-miscible solvent, usually water. In each case, a catalyst, e.g. benzoyl peroxide, is present. Polymerization is catalysed also by heat, irradiation or light.

Emulsion polymerization is cheaper than the solution method and is also amenable to a greater degree of control. The process can be stopped at any stage of the polymerization and, in addition, emulsions can be produced in any desired particle-size range.

Polymers which are to be used in solution form can be separated by filtration after acidification of the emulsion. The product is dried and dissolved in a suitable solvent. Considerable quantities of the polymer emulsions are used directly in the manufacture of latex emulsion paints.

The following are the principal types of vinyl resin.

Polyvinyl acetate (PVA) is a colourless resin with a low softening point (about 38°C, 100°F), non-toxic and thermoplastic. It is supplied in a number of grades differing in degree of polymerization and, to some extent, solubility characteristics. The lower grades are soluble in ethyl alcohol, esters, ketones, and aromatic solvents. They are insoluble in white spirit. Films produced from solutions of the resin tend to be brittle and lack adhesion on smooth surfaces. They are usually plasticized with up to 20 percent ester-type plasticizer such as dibutylphthalate, but gradual loss of plasticizer by volatilization or absorption into a porous substrate will lead ultimately to embrittlement. The plasticized forms possess improved adhesion, but all films of polyvinyl acetate tend to absorb appreciable quantities of moisture.

The polyvinyl acetate chain has the following structure—

$$\left[\begin{array}{c} CH-CH_2-CH-CH_2 \\ | \qquad\qquad | \\ O.COCH_3 \quad O.COCH_3 \end{array} \right]_n$$

Polyvinyl formal. Polyvinyl acetate can be hydrolysed to polyvinyl alcohol and this, in turn, can be reacted to give acetals.

Reaction of polyvinyl alcohol with formaldehyde yields polyvinyl formal:

$$\left[\begin{array}{c} CH-CH_2-CH \\ | \qquad\qquad | \\ O-\!-CH_2-\!-O \end{array}\right]_n$$

In combination with phenol-formaldehyde resins, polyvinyl formal is used in wire enamels.

Polyvinyl butyral. This is produced in a similar way to polyvinyl formal, using butyraldehyde in place of formaldehyde:

$$\left[\begin{array}{c} CH-\!-CH_2-\!-CH \\ | \qquad\qquad\qquad | \\ O-\!-CH-\!-O \\ | \\ (CH_2)_2 \cdot CH_3 \end{array}\right]_n$$

Polyvinyl butyral is used in etch and some types of blast primer. Both butyral and formal resins show very good adhesion to most surfaces.

Polyvinyl chloride (PVC) is produced from vinyl chloride ($CH_2 = CH \cdot Cl$) by one of the recognized methods of polymerization, i.e. bulk, emulsion, or suspension. The operation must be carried out under very close control in view of the recognized carcinogenic properties of vinyl chloride.

It is a colourless polymer with a high Glass Transition Temperature (Tg) of approximately 80°C (176°F). (Tg is the temperature at which a polymer changes from a hard rigid state to a soft plastic condition—see p. 274.)

PVC has very limited solubility in common solvents, and in surface coatings it is used either as a suspension (plastisol or organosol) or as a co-polymer (*below*).

Plastisols consist of PVC dispersed in a suitable plasticizer. They are usually pigmented and yield thick coats which are stoved at 170°–193°C (338°–379·4°F). At this temperature the PVC absorbs the plasticizer and becomes sufficiently soft to form a continuous film. Such films are extremely hard and resistant to most reagents. On bare metal the adhesion is sometimes suspect, and they are often applied over an epoxy primer.

Organosols are dispersions of PVC in plasticizer and thinner. They are pigmented and can give thinner and smoother coats than the plastisols. They require similar stoving treatment.

The solids content of organosols is higher than that of vinyl solutions, and they are replacing these solutions in the internal coating of metal containers [6].

Both organosols and plastisols are used in the coil coating of metal strip.

Vinyl co-polymers

Vinyl acetate can be co-polymerized with a wide range of other unsaturated monomers to give a variety of products with any desired degree of hardness or flexibility. The need for added plasticizer, so essential with the homo-polymer, polyvinyl acetate, is eliminated. Among monomers commonly used are acrylates, maleates, and caprates. Emulsion polymerization is the common method of production.

In the decorative field very considerable quantities of vinyl co-polymer emulsions are used. These have, for a number of years, been based on the co-polymers mentioned above, but a definite improvement was made when ethylene was co-polymerized with vinyl acetate. These co-polymers, containing 8–25 percent of ethylene, have proved to be superior to other co-polymers in durability, resistance to reagents, and non-yellowing properties. An additional advantage, particularly in cold countries, is in film formation, i.e. coalescence of the dispersed polymer. This will take place down to $0°C (32°F)$.

Solutions of vinyl chloride/vinyl acetate and other vinyl co-polymers are used for a variety of coating applications. They show good resistance to light, weather, and chemicals, but since the solutions generally have low solid contents the applied films tend to be thin. Multicoat systems are therefore required to achieve an appreciable film thickness. It is possible to produce "high build" pigmented vinyl paints by the use of thickeners.

Solution type vinyl coatings are used for the protection of chemical and process plant. The vinyl chloride/vinyl acetate co-polymer can be added to certain long-oil alkyd paints to shorten the drying time and to reduce the period before overcoating.

When coated metal plate is subjected to severe deformation during forming it is an advantage to apply a vinyl "size coat" to the metal. It is claimed that this improves the adhesion of finishes based on acrylics or polyesters [7].

POLYSTYRENE RESINS

Styrene or vinyl benzene, $C_6H_5CH = CH_2$, is the basic monomer for this type of resin. It is manufactured by reacting benzene and ethylene at high temperature and pressure in presence of a catalyst. Ethyl-benzene is produced which is then dehydrogenated to yield styrene:

$$C_6H_6 + CH_2 = CH_2 \rightarrow C_6H_5CH_2CH_3$$
$$C_6H_5CH_2CH_3 \rightarrow C_6H_5CH = CH_2 + H_2$$

The product is then fractionally distilled in vacuum, sulphur being present to prevent polymerization of the styrene. This and other methods of synthesis, as well as the properties and polymerization of styrene, have been described in the literature [8].

Vinyl toluene, $CH_3C_6H_4CH=CH_2$, which consists of a mixture of the meta and para isomers, is manufactured in a similar manner. During the preparation a proportion of the ortho isomer is produced. This is separated by fractional distillation under vacuum.

Properties of styrene

It is a mobile liquid with a characteristic odour and polymerizes readily. Premature polymerization is inhibited by addition of a very small quantity of p-tertiary butyl catechol or hydroquinone.

Polymerization of styrene

Polymers of styrene itself are of little interest to the paint industry, but styrene forms a number of interesting and useful co-polymers. The co-polymerization reaction generally requires free-radical initiation and can be carried out either in solution (in aromatic hydrocarbon or carbon tetra-chloride) or in emulsion. The exclusion of air is advisable, since oxygen retards the reaction.

Emulsion polymerization is the most widely used method, the free radical initiation being effected by either a redox system or a persulphate.

Co-polymers of styrene

The styrene–butadiene co-polymer has been used for a number of years as the basis of a decorative emulsion paint in U.S.A. but has not gained acceptance in the U.K. [9].

Co-polymers of styrene and acrylic acid derivatives form the basis of many thermosetting acrylic systems (see page 233).

The co-polymers of vinyl toluene and acrylic esters, in solution form, are used as binders in exterior masonry finishes. They show good durability. Styrene becomes an integral part of polyester finishes which are described on page 218.

One of the most interesting co-polymerization reactions of styrene and, more particularly, vinyl toluene, is with alkyd resins. These are described on page 217.

ACRYLIC RESINS

These are closely related to the vinyl resins. The acids acrylic, $CH_2=CH.COOH$ and methacrylic $CH_2=C(CH_3)COOH$ as well as their esters can be polymerized to form long-chain thermoplastic resins:

$$CH_2=\overset{\displaystyle COOCH_3}{\underset{\displaystyle CH_3}{C}} \quad \rightarrow \quad -CH_2-\overset{\displaystyle COOCH_3}{\underset{\displaystyle CH_3}{C}}-CH_2-\overset{\displaystyle COOCH_3}{\underset{\displaystyle CH_3}{C}}-CH_2-$$

Methyl methacrylate   **Polymethyl methacrylate**

The most interesting and useful polymers are obtained by co-polymerization, whereby monomers can be chosen to impart specific properties to the film. Polystyrene gives brittle films which require addition of plasticizer, but styrene can be co-polymerized with one or more acrylates to give films of any desired degree of hardness and flexibility. Judicious choice of monomers will give resins which produce colourless, resistant films with good exterior durability.

Polymerization takes place readily and is catalysed by heat, light or oxidizing agents such as peroxides, ozone and oxygen. The reaction can be carried out by either solution or emulsion methods.

Properties

In general, acrylate polymers are colourless, soluble in esters, ketones and aromatic solvents. They are insoluble in aliphatic hydrocarbons and are resistant to oils and greases. As a result of the number of possible monomers, a wide variety of film properties is possible — from rubbery to hard, almost brittle. They possess good lightfastness, adhesion, durability, and electrical properties. The emulsions also possess good adhesion, and when pigmented yield flexible and durable films.

Thermosetting acrylic resins

The acrylic resins described above are long-chain thermoplastic materials and so not suited to stoving procedures. It is, however, possible to introduce reactive groups, spaced along the polymer chain, which can be cross-linked with other groups to give complex three-dimensional structures. Groups introduced include carboxyl, hydroxyl, amino, and epoxy.

Thermosetting acrylic resins are, in general, based on the monomers methacrylic ester, acrylic ester, and styrene. Other monomers are incorporated to provide the reactive groups. The nature and relative amounts of monomers control the flexibility and other properties. For example, methacrylates and acrylates are used for maximum durability. The incorporation of styrene or vinyl toluene yields finishes resistant to detergents and household chemicals, particularly when cross-linked with epoxy resin (see below).

Carboxyl groups can be introduced by the use of a proportion of methacrylic acid. Cross-linking of the product with epoxy resins gives a film of outstanding chemical resistance, hardness, and adhesion.

Hydroxyl groups are introduced by using hydroxyalkyl methacrylates or acrylates. The resulting resin can be cross-linked with hexamethoxymethyl melamine to give films of outstanding weather-resistance and gloss retention. They are used as stoving finishes on cars or can be used as acid catalysed lacquers.

Acrylic resins incorporating acrylamide can be reacted with formaldehyde and butanol giving a butoxymethyl acrylamide monomer, containing the group

$$-CH_2-\overset{|}{C}H-CO-NH\cdot CH_2OC_4H_9$$

This group is similar to that in a butylated amino resin and will cross-link with other materials containing free —OH groups such as alkyds and epoxies. When cross-linked with the latter, the resulting films have outstanding resistance to fats, oils and household chemicals. They are, therefore, used on domestic appliances.

Acrylic systems used for the finishing of furniture are suited to curing by ultraviolet radiation, and the advantages of these coatings have been reviewed by Dufour [10]. While this form of cure has been found to operate successfully with clear coatings, difficulties have been experienced with pigmented coatings, particularly of the high-build type. These difficulties have been examined by Rybuy and Vona [11]. Suggestions for their solution have also been advanced by Parrish [12] and by Davis *et al.* [13] who consider that the solution lies in the type of photoinitiator employed.

SILICONE RESINS

These are semi-inorganic high polymers containing both silicon and carbon, and from these elements it is possible to prepare a large number of silicone polymers according to the number and types of groupings. They can be "tailored" for a great many specific purposes and are marketed as resins (solutions), oils, and rubbers.

Formation of silicones

The chemical reactions involved in the formation of silicone resins are many and complex. The following scheme indicates the general reactions involved:

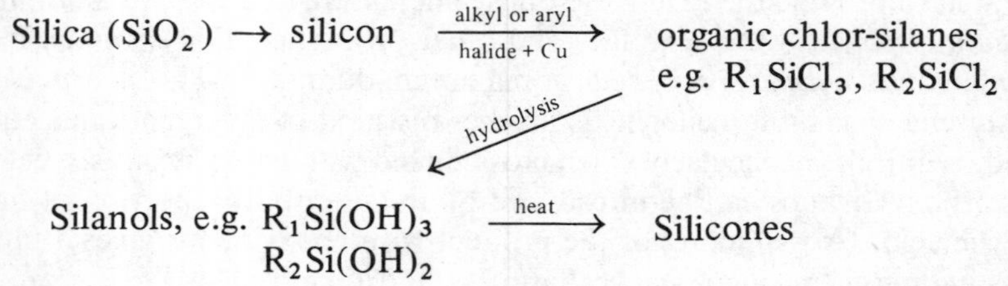

Silica (SiO_2) $\rightarrow$ silicon $\xrightarrow[\text{halide + Cu}]{\text{alkyl or aryl}}$ organic chlor-silanes e.g. R_1SiCl_3, R_2SiCl_2

$\xleftarrow{\text{hydrolysis}}$

Silanols, e.g. $R_1Si(OH)_3$ $\xrightarrow{\text{heat}}$ Silicones
$R_2Si(OH)_2$

Properties

The nature of the alkyl and/or aryl groups determines, to a large degree, the properties of the resin. Methyl silicones are useful as waterproofing agents but find no application as paint materials. The introduction of a phenyl group confers heat resistance, but pure phenyl silicones are too brittle and fuse readily. All-round properties of a high standard are exhibited by co-polymers of methyl and phenylsilicone. These are far superior to the methyl or phenyl resins in respect of heat resistance, flexibility, electric properties, and durability.

Silicone resins are supplied in solutions in aromatic solvents, and can be thinned with similar solvents as well as esters and ketones. They are not soluble in white spirit.

Silicone resins will not cure at ambient temperatures and require heating to 200° to 250°C (392° to 482°F) for about two hours for complete cure. The process can be accelerated by the use of the naphthenates or octoates of zinc, cobalt or iron.

The films produced can vary from rubbery to glass-like hardness. The softer types are the more resistant to heat and weathering, but to obtain the necessary mechanical properties the two types can be blended.

Uses

Silicone resins are used as the sole binder in paints to withstand high temperatures, particularly when pigmented with aluminium. Such paints require a curing period of 2 hours at 200°–220°C (392°–428°F) after which they will withstand prolonged exposure to temperatures up to 430°C (806°F). With other heat-resisting pigments the service temperature is lower, about 280°–300°C (536°–572°F).

Binders consisting of co-polymers of alkyds and silicones, containing up to 30 percent silicone, are now being used for gloss paints with outstanding weather-resistance — see p. 217.

PHENOLIC RESINS

The formation of resinous materials by the reaction between phenol and formaldehyde has been known for many years and was put to commercial use by Baekeland in the early years of this century. He produced the first plastic, "Bakelite", and so stimulated a study of the reaction between various phenols and formaldehyde which has produced many valuable resins.

The reaction between phenol and formaldehyde alone is slow, but takes place rapidly in the presence of acid or alkali catalysts. The two types of catalyst produce different products.

Acid catalyst

When phenol and formaldehyde (formalin) solution are boiled in the presence of 0·5 percent of a strong mineral acid, reaction takes place fairly rapidly and results in the formation of two distinct layers. The upper aqueous layer is run off and the residual layer heated to 150°C (302°F) to remove all volatile matter. On cooling, a hard, brittle, amber-coloured resin is produced. The reaction can be represented thus:

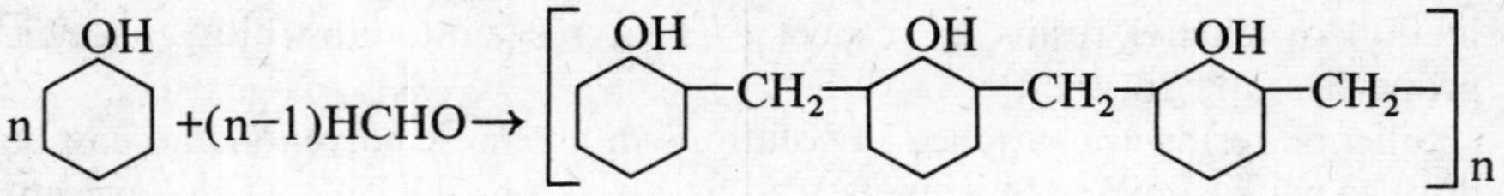

The molar ratio of formaldehyde to phenol is less than one, and any excess phenol is removed in the final heating. Resins produced in this way are known as *novolacs* and are thermoplastic. They are widely used in the plastics industry but find little application in paints as they are insoluble in oils and hydrocarbon solvents.

Alkali catalyst

If phenol and formaldehyde are heated together with the addition of about 1 percent sodium hydroxide or lime, reaction takes place with the formation of a resin of somewhat different constitution from that produced under acid conditions. The reaction can be represented thus:

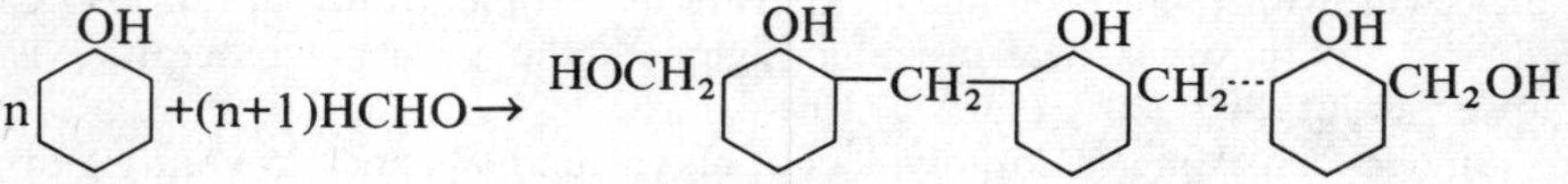

The molar ratio of formaldehyde to phenol is greater than one, and reaction can also take place in the para position. The product contains a range of molecules of differing molecular complexity and this is reflected in considerably variation in viscosity and reactivity.

These resins, known as *resoles*, are brown syrups, soluble in alcohol and are thermosetting, i.e., on further heating they react to produce insoluble and infusible products. They are used in the plastics industry, but their insolubility in oils and hydrocarbons results in little application in the paint industry.

If, however, ammonia is used in the preparation, resins of the following type are produced:

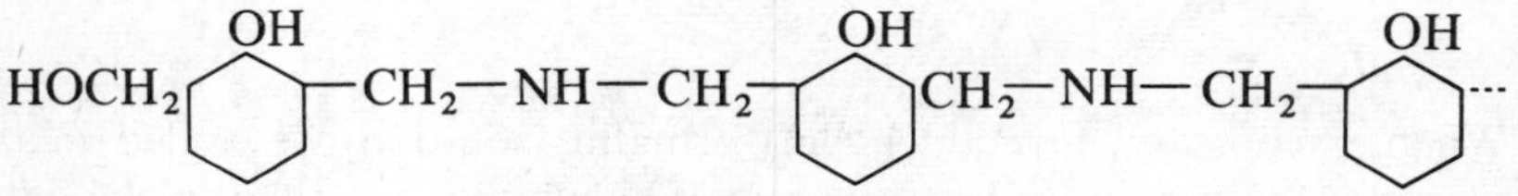

Such resins are sometimes used in stoving lacquers for cans, but their greatest use is in varnish-making as they are soluble in oils.

Modified phenolic resins

The difficulties encountered in the early days in producing oil-soluble phenolic resins led to the examination of their compatibility with other resins. It was found that the resoles reacted with rosin to give products which could be esterified with glycerol, and the resulting resins were soluble in oils. These modified resins usually contain 10 to 25 percent of the resole, and their general properties depend on the type of phenol used. Phenols substituted in the para position by tertiary butyl and octyl groups give very useful varnish resins. The greatest resistance to yellowing is provided by the use of diphenylol propane. All these modified resins possess melting points appreciably higher than that of rosin.

Properties

The rosin-modified phenolic resins are brittle solids with melting points between 100° and 160°C (212° and 320°F). The unesterified types are highly acid (AV 100 to 120) and are soluble in alcohols, hydrocarbons and drying oils. They react rapidly with basic pigments. Esterification yields resins of acid value 15 to 20 which are insoluble in alcohols and which vary in solubility in hydrocarbons and oils. Yellowing on exposure is a general property of these resins, but the least discoloration is shown by resins based on diphenylol propane.

They are used widely in varnish-making in conjunction with linseed, enamel oils, and wood oil. With the latter they retard the gelling tendency and give varnishes with high water-resistance.

Generally, varnishes made on these rosin-modified phenolic resins are used in a wide range of paints including decorative undercoats, primers, and marine paints. They are also used in certain types of printing ink.

Pure oil-soluble phenolic resins

It has been shown that the rate at which phenolic resins, based on substituted phenols, cross-link to form insoluble and infusible products is controlled by the nature and position of the substituent group. Para-substituted phenols react most slowly and, furthermore, the rate decreases with increase in the size of the group. As the groups become large, not only does the rate of reaction fall but the resin develops definite solubility in oils. Such para-substituted phenols can be reacted to give either novolacs or resoles, and these in turn are known as *non oil-reactive* and *oil-reactive* respectively.

Non oil-reactive resins

These are commonly based on *p*-phenyl-phenol and are prepared by the normal acid-catalysed method. They are soluble in aromatic hydrocarbons and possess softening points of about 100°C (212°F). Varnishes are made very easily by heating the resin with the drying oil, and the films produced are durable and possess a high degree of alkali and water resistance. They are used for spar varnishes, insulating varnishes and in machinery paints.

Oil-reactive resins

These are made from *p*-tertiarybutyl and *p*-octyl phenols by reaction with formaldehyde in the presence of an alkali catalyst. Sufficient sodium hydroxide is present to dissolve the phenol, and the resin is subsequently precipitated with acid. It is then washed thoroughly and heated in vacuum to remove all volatile materials.

Properties and uses

The oil-reactive phenolics are soluble in hydrocarbons and soften at 75° to 80°C (167° to 176°F). When heated with oils, reaction occurs through the reactive methylol ($-CH_2OH$) groups. This reaction is vigorous with conjugated oils such as dehydrated castor and wood oils and yields varnishes with a higher degree of resistance than the oil-soluble type. These show very high resistance to water and alkalis and are used as spar varnishes and in chemical-resistant finishes.

These resins also react with rosin and ester gum to produce the rosin-modified phenolic resins.

Cashew nutshell liquid

This is a reddish-brown viscous liquid which consists of approximately 90 percent anacardic acid (a derivative of salicylic acid) and 10 percent cardol (a monohydroxy phenol). On strong heating anacardic acid is decarboxylated and converted into cardol.

The liquid can be polymerized to give resins soluble in both aliphatic and aromatic hydrocarbons and possessing very good resistance to acid and alkali.

MALEIC RESINS

These are produced by reaction between maleic anhydride and rosin, utilizing the Diels–Alder addition reaction. Addition of maleic anhydride to a conjugated unsaturated system takes place according to the general scheme:

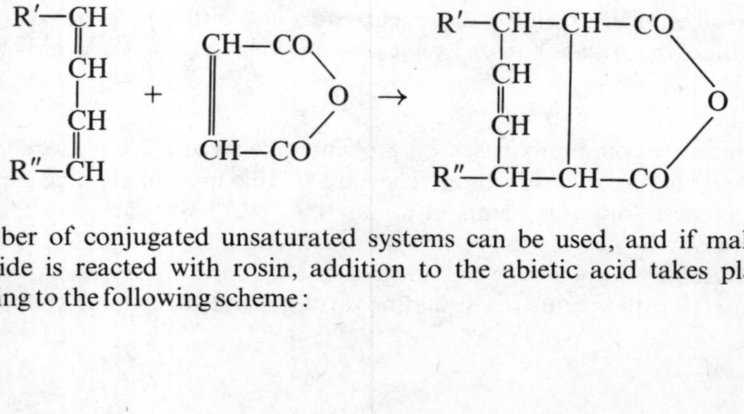

A number of conjugated unsaturated systems can be used, and if maleic anhydride is reacted with rosin, addition to the abietic acid takes place according to the following scheme:

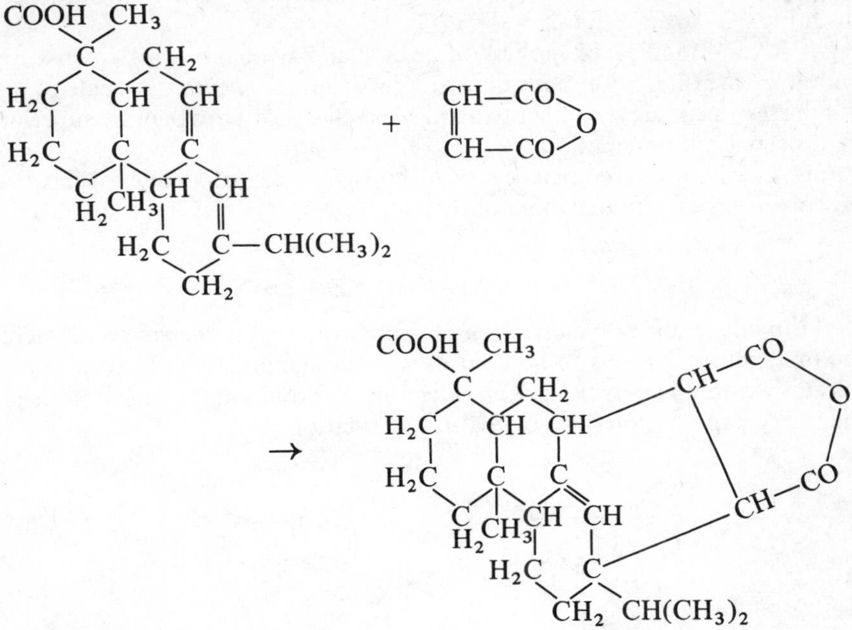

$$\text{COOH } CH_3$$

(rosin–maleic anhydride reaction structures)

$$\begin{array}{c} CH\!-\!CO \\ \| \qquad\;\; O \\ CH\!-\!CO \end{array}$$

$+$

$\rightarrow$

The product can then be esterified with glycerol or pentaerythritol, and in practice the three ingredients, rosin, maleic anhydride and polyhydric alcohol are sometimes heated together until reaction is complete.

Properties and uses

The properties of these resins depend on the amount of maleic anhydride used, the type of rosin, and type of polyhydric alcohol. Not more than 12 percent of maleic anhydride is used, as resins approaching this composition have high melting points and low solubility. Pentaerythritol gives harder films than glycerol.

Maleic resins are soluble in hydrocarbon solvents, esters, ketones and drying oils. Their ready solubility renders them suitable for *cold-cut* varnish-making, but they are often dissolved in oils by heating. They give pale varnishes, and with linoleic-rich oils produce non-yellowing whites and tints. They are also used in stoving lacquers for metal decorating.

The chemical and alkali resistance of the films is poor, but for normal exposure, when they are frequently blended with alkyds, the performance is satisfactory. They are often added to cellulose nitrate lacquers to improve gloss and adhesion.

Maleic resins bear some resemblances to three other types of resin:

(i) Ester gum. The maleics behave as ester gums which have been hardened by the maleic–rosin adduct.

(ii) Rosin-modified phenolics. Maleics blend with oils with equal ease and give varnishes with many similar properties. However, the maleics do not possess the same resistance to alkalis and chemicals, but they are superior in respect of colour retention.

(iii) Alkyds. The resemblance here is through the use of a polybasic acid, a polyol and an unsaturated monobasic acid.

COUMARONE–INDENE RESINS

Resins made by the polymerization of coumarone and indene mixtures were the first synthetic resins to be used in varnish manufacture. They are produced by catalytic polymerization of coumarone and indene which occur in the heavy naphtha fractions of coal-tar distillation.

Coumarone   Indene

Polymerization takes place through the double bond and is usually catalysed by concentrated sulphuric acid, but other catalysts are sometimes employed. The formation of polymers from coumarone can be represented in the following scheme, and that from indene is similar:

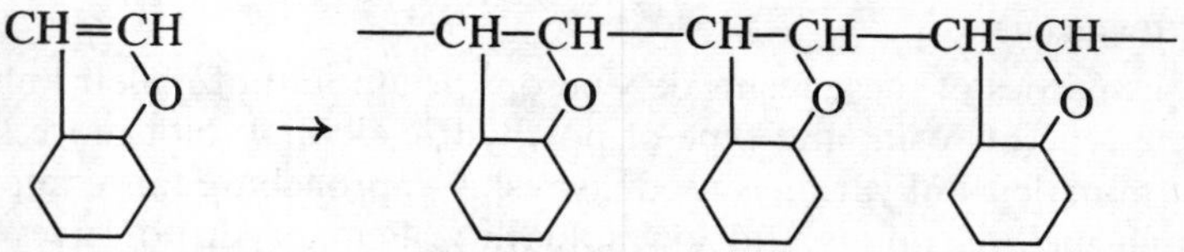

The indene is sometimes separated and converted to polyindene resin, but commercial coumarone resins are usually a mixture of the two.

These resins are marketed in a range of colours from pale amber to dark brown and from soft gums to brittle solids. The melting points range from 90°C to about 150°C (194° to 302°F) according to the degree of polymerization. They yellow on exposure to sunlight, are very inert and resemble natural resins in a number of properties. Thus, they are insoluble in alcohols and soluble in esters, acetone, and aromatic hydrocarbons. The solubility in aliphatic hydrocarbons varies from partial to complete according to grade.

Coumarone–indene resins are soluble in drying oils and can be used in cold-cut varnishes or can be dissolved in oils by heating. Such varnishes have very low acid values (< 1) and so are used in the manufacture of metallic paints such as aluminium and bronze, where the brilliance and leafing properties of the metals are destroyed by small amounts of free acid.

Varnishes incorporating coumarone–indene resin and wood oil show high resistance to attack by alkali and are therefore used in primers for alkaline plasters, asbestos, and cement render.

POLYIMIDE RESINS

These resins are of particular interest by reason of their outstanding heat-resisting properties. They are formed by a condensation polymerization of the dianhydride of a tetracarboxylic acid with an aromatic diamine:

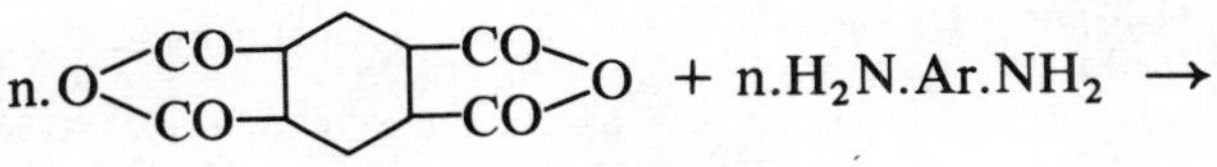

Pyromellitic anhydride

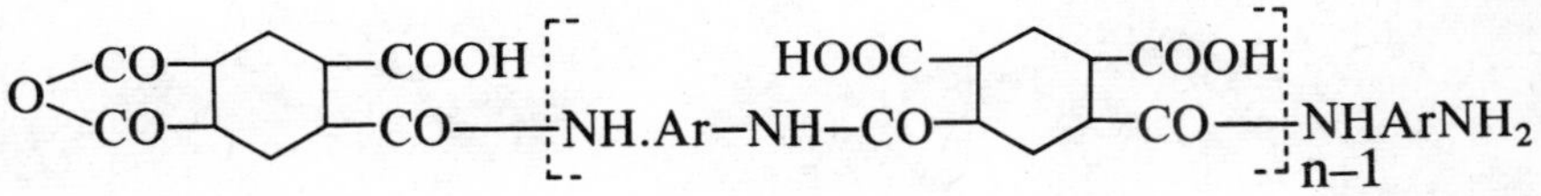

Polyamic acid

A solution of polyamic acid is applied to the article, dried and then stoved at a temperature in excess of 150°C (302°F). Dehydration takes place with formation of polyimide resin:

Polyamic acid →

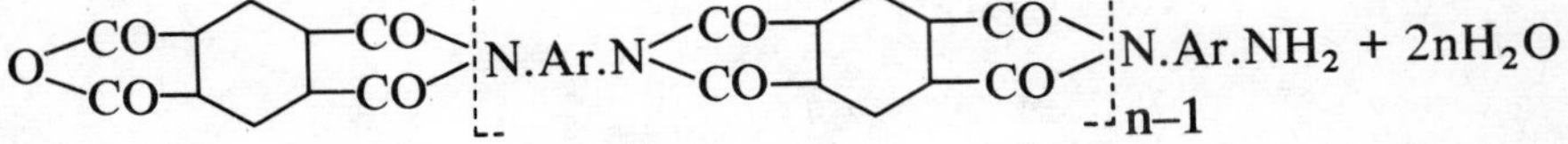

Polyimide resin

By the use of different dianhydrides and diamines a range of resins can be produced. Many of these are stable for long periods at 300°C (572°F) and at higher temperatures for shorter periods [14].

References to Chapter 13

[1] KIENLE, R. H., *JSCI*, 1936, **55**, 229T.
[2] MARK, H. & WHITBY, G. S. (eds), The collected papers of W. H. Carothers on high polymeric substances. Interscience, New York, 1950.
[3] HOLMBERG, K., The rôle of the Diels–Alder reaction in the curing of drying oil alkyd–melamine systems. *JOCCA*, 1978, **61**, 356.
[4] *Idem*, The mechanism of the acid catalysed curing of alkyd–melamine systems. *Ibid.*, 359.
[5] BRUNNER, H., in Hess' *Paint Film Defects*, 3rd edn. Chapman & Hall, London, 1979.
[6] REED, R. T., Recent developments in protective finishes for metal containers. Part 1. Recent developments in internal protective finishes. *JOCCA*, 1975, **58**, 51.
[7] HOLT, J. C., Recent developments in protective finishes for metal containers. Part 2. External organic finishes. *JOCCA*, 1975, **58**, 57.
[8] WARSON, H., *POCJ*, 1971, Aug. 13, 163.
[9] *Idem*, *PPCJ*, 1971, Sept. 17, 259.
[10] DUFOUR, P., UV cured acrylic coatings for wood. *JOCCA*, 1979, **62**, 59.
[11] RYBUY, C. B. & VONA, J. A., New developments in ultraviolet curable coatings technology. *JOCCA*, 1978, **61**, 179.
[12] PARRISH, M. A., Ultraviolet light curing: some benefits and recent advances. *JOCCA*, 1977, **60**, 474.
[13] DAVIS, M. J. *et al.*, The UV curing behaviour of some photoinitiators and photoactivators. *JOCCA*, 1978, **61**, 256.
[14] BAWN, C. E. H., Polymers: developments for the future. *JOCCA*, 1973, **56**, 423.

14 Rubbers, bitumens, pitches, gums and glues

RUBBER DERIVATIVES

Rubber is obtained in the form of an aqueous emulsion, known as "latex", by tapping trees of the species *Hevea braziliensis* which grows in Malaya, Indonesia and Brazil. The milky liquid is treated with acetic or formic acid to coagulate the rubber, and the resulting spongy mass is rolled and washed to produce white crepe rubber. Other methods of treatment are employed but give products of inferior colour. The white crepe rubber is preferred for the rubber derivatives used in surface coatings.

Chlorinated rubber

The raw rubber molecule contains a number of unsaturated linkages and will react with chlorine to yield a solid product known as chlorinated rubber. Synthetic rubber reacts in a similar way and is used in the process. Both addition and substitution reactions take place, the latter resulting in the formation of hydrochloric acid. In manufacture, the rubber is dissolved in carbon tetrachloride in an acid-resisting vessel fitted with means of heating and cooling. Chlorine is passed into the solution, and on completion of the reaction the solution is washed free from hydrochloric acid with aqueous alkali. Finally the chlorinated rubber is precipitated by addition of methanol, washed and dried. It is marketed as a white powder containing 65–68 percent chlorine [1]. A range of viscosity grades is available, the viscosity being designated by numbers [2]. These represent the viscosity, in centipoises, of a 20 percent w/w solution in toluene.

Properties and uses

It is difficult to remove carbon tetrachloride entirely from chlorinated rubber and the powders usually contain about 3 percent of the solvent [1], a fact which should not be overlooked when large quantities are being used.

Solvents

It is soluble in aromatic hydrocarbons, chlorinated hydrocarbons, esters, and some ketones. Blends of solvents are usually employed, the composition depending on the method of application of the paint. The use of a proportion of a polar solvent is necessary to prevent "cobwebbing" in application by conventional spray.

Chlorinated rubber is a non-convertible binder and, as is common with this class of material, difficulties can be encountered in applying a second coat by brush. The solvent in the latter softens the first coat and "pick-up" or "lifting" occurs. The inclusion of a proportion of non-solvent such as white spirit reduces this pick-up very considerably.

Heat stability

This is limited and chlorinated rubber coatings should not be used on surfaces likely to exceed 80°–85°C (176°–185°F) for any length of time. Liberation of hydrochloric acid is possible, and it is usual to include a "stabilizing" agent to absorb the acid. Certain epoxy compounds such as epoxidized soya bean oil are used.

General resistance properties

Chlorinated rubber possesses an extremely low moisture permeability and is therefore used on surfaces which may be permanently in contact with water, such as the submerged areas of swimming pools.

It is resistant to alkali, dilute acids, mineral oils and greases, inorganic chemicals, aliphatic hydrocarbons, and alcohols. It is not resistant to aromatic hydrocarbons, certain esters and ketones, chlorinated solvents, animal oils and fats.

The durability is of a high order and chlorinated rubber is used on steel structures where a life of 10–15 years is required before major repainting.

Compatibility

Chlorinated rubber is compatible with a wide range of other resins, including alkyds, acrylics, hydrocarbon resins and rosin. Blends with polyurethanes are used as adhesives.

It is often blended with certain medium- and long-oil alkyds to upgrade chemical resistance, hardness, and resistance to lubricating oils. The drying times also are reduced. Blends of this nature are used on ships' topsides and superstructures and in primers for chlorinated rubber systems on steelwork.

Blends with certain acrylic resins give tough films with high resistance to chalking under intense solar radiation [1]. They also show good adhesion to galvanized iron and etched zinc surfaces.

Plasticizers

A chlorinated rubber film deposited on evaporation of solvent from a solution is brittle and lacks adhesion. It is therefore plasticized, the amount of plasticizer varying from 20 to 50 percent according to the degree of pigmentation and the end-use of the paint. Plasticizers employed include chlorinated paraffin waxes, many esters including phthalates, phosphates, adipates, sebacates and some drying oils. The most widely used are the chemically inert chlorinated paraffin waxes [3].

Hazards

These are usually associated with the solvents. As mentioned earlier, the chlorinated rubber powder generally contains about 3 percent carbon tetrachloride.

Cyclized or isomerized rubber

When raw rubber is heated with phenol and a catalyst, a change in structure takes place, resulting in a less unsaturated product with an iodine value of 50 to 100.

It is a tough solid material, soluble in aliphatic hydrocarbons and available in a number of viscosity grades. It is compatible with drying oils, varnishes, long-oil alkyds and some medium-oil alkyds. It is not compatible with short-oil alkyds.

Uses

When used alone as a film-former, cyclized rubber requires a plasticizer and is readily compatible with the chlorinated paraffin waxes. Such films require a cobalt drier and show a high degree of chemical resistance, but the resistance to solvents is poor.

The wide range of compatibility of cyclized rubber enables it to be used to upgrade the chemical and water resistance of long-oil alkyds and varnishes, the composition of the mixture being determined by the severity of the conditions. Mixtures with alkyds or varnishes dry to films which show a greater solvent resistance than the cyclized rubber, and these can be over-painted without difficulty.

The general chemical resistance and exterior durability of cyclized rubber is not as high as with chlorinated rubber.

BITUMINOUS MATERIALS

These are generally black or brownish-black materials consisting mainly of hydrocarbons, and soluble in carbon disulphide and certain other solvents of a neutral character. They contain a small amount of free carbon and materials known as asphaltenes to which the colour is largely due.

Bitumens are obtained from natural sources (in which they often occur in association with mineral matter) or as residues from the distillation of crude petroleum oils.

NATURAL BITUMENS

Bitumen occurs in admixture with mineral matter, the proportion of which varies from one location to another. Where the mineral content is high, as in the Trinidad Lake, the product is known as asphalt and is used for road-making. The term asphalt, however, is used rather loosely and is often applied to other types. Materials in which the bitumen predominates and in which the mineral matter content is often very low are of interest for coatings. The following are the most important types.

Gilsonite. This is one of the most widely used materials of the class and is obtained from North America. It is a lustrous jet-black brittle solid which gives a brown streak when rubbed on paper and breaks down to a brown powder. When cold it breaks with a conchoidal fracture but softens progressively on heating, finally becoming liquid at about 170°C (338°F). The specific gravity varies between 1·05 and 1·10. It contains very small amounts of mineral matter and is soluble in petroleum solvents, coal-tar solvents and carbon disulphide. Gilsonite is compatible with most of the oils and resins used in varnish manufacture and is used in stoving blacks, black japans and some dipping blacks.

Manjak or glance pitch. This is similar in appearance to gilsonite but gives a black streak on paper. It is obtained from both Barbados and Trinidad, the latter type being preferred in view of greater staining power. Trinidad Manjak softens at about 170°C (338°F) but differs from gilsonite in showing limited compatibility with oils. Films produced from Manjak possess good weathering properties.

Grahamite is mined in the southern states of the U.S.A. and in Mexico. It is dense black, giving a black streak on paper and is harder and less fusible than gilsonite (softening point 200° to 300°C, 392° to 572°F).

Rafaelite is obtained from the Argentine and is harder than gilsonite. It also possesses a higher specific gravity. The appearance is lustrous black and it breaks with a conchoidal fracture. The melting point is above 200°C (392°F) and the compatibility with drying oils very limited.

PETROLEUM BITUMENS

Residual bitumen is obtained from the residues obtained in the distillation of crude asphaltic petroleums. The properties vary with the type of crude material and the conditions of distillation. However, conditions of processing can be closely controlled to give a uniform product from a particular type of crude. They exhibit a wide range of softening points and are very susceptible to changes in temperature, becoming sticky in warm weather.

Blown or oxidized bitumens are prepared by blowing air through semi-liquid petroleum residues at about 280° to 300°C (536° to 572°F). The products are semi-liquids or solids of melting point up to 120°C (248°F). They are less susceptible to temperature changes than the unblown material but have very limited compatibility with the common varnish-making materials.

Petroleum resins or Albino bitumens. These are obtained from petroleum oils free from asphaltenes and vary considerably in colour. Many specimens are similar in colour to dark rosin. They are soluble in hydrocarbons and are compatible with drying oils.

General properties of bitumens

The intensely dark colour of bitumens is useful in black finishes but limits their use in coloured paints.

They are very resistant to water penetration and to attack by acids and alkalis. They are, however, very sensitive to changes in temperature and some types soften considerably on slight warming.

When exposed to the air, bitumen films undergo no major chemical change, but oxidation of the surface takes place slowly. The product is harder than the underlying bitumen and tends to craze when changes of temperature occur.

Bitumens are soluble in a wide range of solvents and oils and application of a conventional paint film over one of bitumen will result in the bitumen bleeding into the second coat.

BITUMINOUS PAINTS

This term is used for the products obtained from both petroleum and coal-tar sources, but the petroleum products are the more widely used.

Bituminous paints are designed to take full advantage of the water and chemical resistance of the material and are used as protective coatings on steel and other surfaces where protection is of greater importance than appearance. It is usually necessary to use a blend of bitumens to achieve the desired film properties and these are melted together and then thinned out, usually with aromatic solvents. Drying oils are sometimes incorporated and, in this event, the appropriate quantity of drier is added.

The black colour can be modified by pigmentation, but the range of colours obtainable is limited to greys, dark reds, browns and greens. If petroleum resins are used in place of bitumen, paler shades can be produced but "clean" colours are not possible.

Dipping Black is usually a 50 percent solution in white spirit of a mixture of gilsonite and residual bitumen. The ratio of these can be varied at will to produce any desired degree of film hardness.

Black varnish consists of a 65 percent solution of a medium-soft coal-tar pitch in heavy naphtha.

Brunswick Black is a quick-drying material consisting of a mixture of gilsonite and rosin dissolved in naphtha. Drying oils are sometimes incorporated.

Black japans. Air-drying black japans contain a hard resin such as a fused Congo copal in addition to bitumen and oil. They dry to hard lustrous films. Stoving black japans contain gilsonite, varnish, and possibly stand oil.

The time and temperature of stoving and the consequent film properties can be changed by altering the nature and proportions of the ingredients.

Bituminous aluminium paint. Bituminous paints are used widely to protect metal — chiefly galvanized iron — roofs, but with black paints a considerable amount of the radiation is absorbed. If aluminium paste is mixed with the bituminous paint at the rate of about 400 grams of paste per litre (4 lb of paste per gallon) of product, the aluminium *leafs* in the film, producing a bright metallic lustre with a brownish tint. This serves to reflect a high percentage of the radiation and the building is cooler in consequence.

PITCHES

These are fusible solid or semi-solid materials obtained in the distillation of tars from various sources. The source is usually denoted by the prefix.

Coal-tar pitch is a relatively cheap product obtained from the distillation of coal-tar, and the properties vary with the composition of the tar employed. This in turn varies with the type of coal, temperature of distillation, and type of distillation plant.

Coal-tar pitches are compatible with aromatic solvents but are insoluble in aliphatic hydrocarbons, drying oils, and many natural resins. In general they show pronounced acidity.

Certain grades are compatible with epoxy and polyurethane resins and these combinations are widely used as protective coatings for steelwork under highly corrosive or submerged conditions.

Water-gas pitch or **oil-tar pitch** is derived from tars produced in petroleum distillation. The properties of the product vary with the temperature of distillation, the higher temperatures yielding products similar to some coal-tar pitches. These materials can give coatings of good durability.

Stearine or **fatty acid pitches** comprise the residues obtained in the steam distillation of fatty acids. The solubility and properties vary according to the nature of the fatty acids. Generally they have low melting points (below 100°C, 212°F) and poor opacity. They are compatible with some of the harder bituminous materials and are sometimes used as softeners. Stearine pitches are sometimes used in black stoving enamels as they harden on heating.

GUMS

The gums are organic materials similar in appearance and origin to the natural resins but differing from them in solubility characteristics and composition. Gums are carbohydrates, soluble in water and insoluble in organic solvents.

Gum arabic is imported mainly from North Africa where it is obtained as an exudation from species of Acacia, the most important being *Acacia Senegal*. It usually contains about 4 percent of impurities and has a specific gravity of 1·35 to 1·50. Gum arabic is soluble in water but is precipitated on addition of alcohol or borax. Addition of aldehydes or phenols renders the gum insoluble in water.

The aqueous solution is neutral but liable to attack by bacteria, and therefore requires the addition of a preservative such as 1 per cent of *p*-chloro-*m*-cresol. Gum arabic is used as an adhesive and a binder in some types of artists' colours. The aqueous solution gives no coloration with iodine solution, a test which serves to distinguish it from dextrin (*below*).

Senegal gum is obtained from *Acacia Senegal* and is less friable and tougher than Gum arabic. It is used widely in artists' water-colours.

Gum tragacanth is obtained as an exudation of the species Astragalus, a shrub which grows wild in Asia Minor. The exuded gum dries on the surface of the stem and forms characteristic horny nodules. It swells to a gel in cold water but goes into solution on addition of alkali. It is used as a binder for artists' water-colours.

Dextrin or **British gum** is derived from starch by treatment with acid or heating. Flour is moistened with a mixture of dilute hydrochloric and nitric acids and heated to 125°C (257°F), or heated alone to 180° to 200°C (356° to 392°F). The colour of commercial dextrin varies from yellow to brown according to the treatment. It possesses good adhesive properties and is often used as a substitute for Gum arabic. Aqueous solutions of dextrin give a deep blue colour on addition of iodine solution.

Farina is similar in nature to dextrin and is prepared from potato starch. It is used in the preparation of adhesives.

GLUES

The materials in this class have the adhesive properties of gums but differ from them in origin and composition. They belong to the class of complex nitrogenous substances known as proteins and are obtained from animal sources. Before the oil-bound distempers and glue-bound ceiling whites were displaced by emulsion paints, considerable quantities of glues and casein were consumed by the paint industry. Now the quantity used is extremely small, but they merit brief mention.

Gelatin is the term applied to the finest quality glue and is contained in the skins, tissues and bones of animals. The commercial materials are sometimes known as "Technical Gelatins". The bulk of the materials are known as glues and are classified according to origin.

Skin glues are manufactured by first heating the skins with lime in baths of increasing strength to remove hair and blood. The skins are then washed thoroughly and extracted with water at progressively higher temperatures and finally in an autoclave at 100° to 110°C (212° to 230°F). The first extracts are the purest gelatin, and with successive extracts the materials become less pure.

Rabbit skin glue is the best known of this type and gives firm gels at low concentrations.

Bone glues are prepared from degreased bones which are treated with cold dilute hydrochloric acid to remove calcium phosphate and other mineral constituents. The residue is washed and then extracted with hot water in a similar way to skin glue.

Size is a weak glue which contains a large proportion of a similar substance, chondrin, derived from cartilage, and which does not form a firm gel. Size forms a very soft jelly and is occasionally used for sealing porous surfaces.

Properties of glues

The best grades of glue are colourless, tasteless and odourless. The specific gravity is about 1·3. Glue swells in cold water and dissolves in hot water, the solution on cooling setting to a gel. The firmness or strength of the gel is taken as a measure of the quality of the glue and is expressed in Bloom values. These are high for stiff gels and low for the softer types.

Glue solutions are very prone to decomposition by bacteria unless treated with an efficient preservative. Materials used for this purpose include p-chloro-m-cresol and sodium pentachlorphenate.

Casein is a protein present in skimmed milk from which it can be precipitated by addition of dilute mineral acid. The product is known as "acid casein". This material swells but does not dissolve in water; it dissolves in presence of an alkali such as ammonia or borax.

Casein is occasionally used as a protective colloid in the styrene–butadiene type of emulsion paints, where it induces good mechanical stability and excellent levelling properties.

Like glue, casein is readily attacked by bacteria unless it is adequately protected by an efficient bactericide.

References to Chapter 14

[1] "Alloprene", Booklet by ICI (Mond Division).
[2] "Alloprene" 5, 10, 20 (ICI).
[3] "Cereclors" (ICI).

15   Cellulose ester and ether products

Cellulose comprises the bulk of the fibrous tissue of all plants, and the purest form is obtained from cotton in which it forms a protective fibre round the seeds. It is a member of the carbohydrate class and the composition of the *repeating unit* can be represented by the formula

$$C_6H_7O_2(OH)_3 \quad \text{or} \quad \left[-O-\begin{array}{c} CH_2OH \\ | \\ CH-O \\ CH \qquad CH- \\ CH-CH \\ | \quad | \\ OH \quad OH \end{array} \right]$$

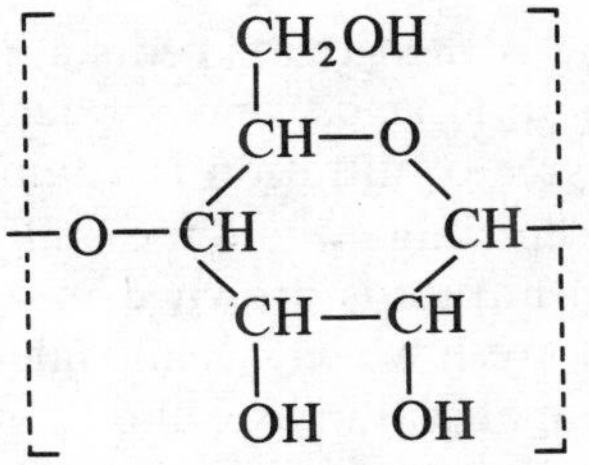

but the actual molecular weight is very high. The unit depicted above contains three alcohol groups, one of which is primary and the remaining two are secondary. The range of esters and ethers derived from cellulose is produced by reactions of these hydroxyl groups, and it is theoretically possible to produce mono-, di- and tri-derivatives. In practice, mixtures are usually obtained.

Esters of cellulose have been known for a considerable time and in 1855 a patent was taken out by Parkes in which he suggested the use of nitrocellulose solutions as protective coatings. He later suggested the use of camphor as a plasticizer and this led to Hyatt's introduction of celluloid in 1869.

At first the use of cellulose esters as paint media was severely restricted by lack of solvents of suitable evaporation rate. The introduction of amyl acetate in 1882 provided a solvent of reasonably slow evaporation, but the high viscosity of the solutions necessitated considerable dilution before application, resulting in very thin films. In addition the plasticizer, camphor, slowly volatilized from the film, leading to embrittlement and breakdown. The production of low-viscosity cellulose nitrate and the introduction of a wider range of solvents and plasticizers led to the mass-production of cellulose nitrate lacquers, and over the last fifty years very large quantities have been produced.

251

CELLULOSE ESTERS

Cellulose nitrate

This is very often—but erroneously—called "nitrocellulose". It was first produced by Schoenbein in 1845, who treated cellulose with a mixture of nitric and sulphuric acids.

The modern method of manufacture employs the same reaction, the sulphuric acid acting both as a dehydrating agent to absorb the water produced and as a catalyst. The cellulose used is either cotton linters, which are short fibres attached to the seed and remaining after the longer fibres have been removed for spinning, or purified wood-pulp. The linters are subjected to preliminary treatments to remove fragments of seed and other impurities and are often given a mechanical treatment to open out the fibres before nitrating.

The cellulose is then treated in stainless steel vessels with a mixture of nitric and sulphuric acids, and the temperature controlled carefully until the desired degree of nitration is attained. The mixture is then run into a stainless steel centrifuge where the bulk of the acid liquor is separated. The crude cellulose nitrate is drowned in water and then subjected to several washings with fresh water. Removal of acid impurities is essential for stability of the material in use, and after washing with water it is heated in slightly acidified water to about 140°C (284°F). This serves to release any sulphuric acid entrapped in the fibres and also reduces the viscosity of the product. The cellulose nitrate is then washed with hot water and centrifuged.

The material still contains about 30 percent water but cannot be dried owing to the dangerous nature of the dry compound. It is therefore "dehydrated" by replacing the water by methylated spirit, butanol or isopropanol. The replacement of water by alcohol is carried out either by forcing the alcohol through the wet cellulose nitrate cake in a press or by spraying the material with the alcohol while it is being spun in a centrifuge.

The use of the higher alcohols has the advantage of more complete dehydration as they are less hygroscopic than the lower members. Nevertheless the material is never completely anhydrous. Cellulose nitrate then reaches the user "damped" with one of the above alcohols; it should not be allowed to dry out after the drums are opened.

Properties of cellulose nitrate

The degree of nitration depends on conditions, and cellulose nitrates are divided into three classes according to the nitrogen content:

> 10·5–11·2% Low nitrogen (L)
> 11·2–11·8% Medium nitrogen (M)
> 11·8–12·2% High Nitrogen (H)

The nitrogen content has an important bearing on the choice of a cellulose nitrate for a given purpose. The most widely used are those with a high nitrogen content, whilst the low-nitrogen grades tolerate large amounts of alcohol and are used in paper lacquers.

The viscosity of a cellulose nitrate solution is of great importance and is influenced by the type of material and the nature of the solvent blend. It is generally measured in the form of a standard solution in aqueous acetone, the solvent consisting of 95 percent acetone and 5 percent water by volume. The concentration of the solution varies with the grade of cellulose nitrate being tested, and viscosity ranges are designated by letters H (high), M (medium), L (low) and X (extra low). The accepted method of designating a cellulose nitrate gives information on type, nitrogen content and viscosity. For example, a material designated DHX 30/50 is a dense (D) type with high nitrogen content (H) and extra low viscosity (X) lying between 30 and 50 poise. (For a full description of the various types of cellulose nitrate and their solution characteristics the reader is referred to Paint Technology Manuals (OCCA), Part 1 [1].)

Testing

The quality of cellulose nitrate must be carefully controlled to ensure successful results. The following tests are among the most important.

Nitrogen content

The older method consisted of measurement of the volume of nitrogen evolved when a known weight of the material is decomposed with concentrated sulphuric acid in a Lunge nitrometer. In the modern method the cellulose nitrate is hydrolysed by sulphuric acid in glacial acetic acid solution. The liberated nitric acid is estimated by titration electrometrically with standard ferrous ammonium sulphate solution in sulphuric acid.

Viscosity

The solution of standard concentration in aqueous acetone is placed in a wide vertical tube in a thermostatically controlled bath at 20°C (68°F). The viscosity is then determined by the falling sphere method (page 178).

Stability

At normal temperatures a properly prepared cellulose nitrate is stable, or at least decomposes at a rate than can be ignored. However, any slight tendency to instability, i.e. decomposition, is accelerated by heat, and resistance to heat is therefore used to assess stability. It is usually carried out by the Bergmann and Junk test in which the cellulose nitrate is heated for 2 hours at 132°C (270°F) and the acid decomposition products are absorbed in water. These are estimated by titration.

Volatile content

When lacquers are prepared from "damped" cellulose nitrate the alcohol content must be taken into consideration. It is usually about 30 percent but is likely to vary. The figure is determined by heating a known weight of the damp material in an aluminium dish in an oven at 60° to 65°C (140° to 149°F). Heating is continued until the weight is constant, and high temperatures must be avoided. The dry material should be re-damped as soon as the estimation is completed.

CELLULOSE LACQUERS

Cellulose lacquers are not simply solutions of cellulose derivatives, but also contain plasticizers to give the necessary film flexibility, and resins to increase the gloss and adhesion. They may also contain non-solvents or diluents to control the cost. A large proportion of the lacquers used is pigmented.

The performance is controlled by the type of cellulose derivative used and by the cellulose–plasticizer–resin ratio. The large number of materials available renders it possible to design cellulose lacquers for a very wide range of uses. Some of the general principles involved in lacquer formulation are discussed below.

Cellulose nitrate lacquers

Cellulose nitrate is the principal component of the film in most cases and there is a wide variety of types available. A low-viscosity type permits the formulation of high-solids lacquers leading to high-build films. Higher viscosity cellulose nitrate gives lacquers of lower solids at spraying viscosity, but the solids can be increased by the use of suitable resins. These lead to greater gloss and hardness of the film, and the types used are discussed below.

Ethyl cellulose is compatible with cellulose nitrate in all proportions and can be incorporated to reduce flammability and increase flexibility.

Resins. These are used to improve the gloss and adhesion of the film. The film produced from cellulose nitrate alone is a semi-gloss and has poor adhesion. The resin must have good compatibility with cellulose nitrate and plasticizer in all proportions, and the amount used varies considerably. Certain cheap resins with good compatibility are used in large proportions in the cheaper types of lacquer, but these are brittle and lack durability.

Damars. These are extensively used on account of good solubility, the dewaxed type being the most suitable. They have a pale colour and confer good gloss and adhesion. Damars are insoluble in alcohol and so cannot be used in the low-nitrogen type spirit lacquers.

Sandarac and **mastic** can be regarded as resins of the same type as damar but have the advantage of ready solubility in alcohol.

Elemi is readily soluble but soft. It is on the borderline between a resin and a plasticizer.

Dewaxed shellac is used in the spirit type of lacquer where it confers good adhesion, toughness and flexibility. Such lacquers are not suitable for exterior exposure.

Ester gums. The glycerol types are completely compatible with cellulose nitrate and soluble in the solvents used. The pentaerythritol types are much less compatible. Low acid value in the ester gum is essential, and the plasticizer content is very important. Generally the proportion of ester gum is high for interior finishes and low in finishes for exterior exposure.

Maleic resins show good compatibility and produce very hard films. These must be well plasticized to avoid brittleness. They are not recommended in lacquers for exterior exposure.

Cyclohexanone resins possess no acidity and are useful in clear lacquers for metals. They have good compatibility and confer good gloss and adhesion on the film.

Alkyds have replaced a number of the natural resins such as kauri, Congo copal, and rosin. Both drying and non-drying types are used, but the latter find widest use. The addition of a non-drying alkyd plasticizes the film and confers good gloss and durability. The films are also resistant to alcohol and to heat.

The castor-oil modified types are used in greatest quantity, but other non-drying types are used for special purposes. The general requirements of such an alkyd are stated [1] to be: fatty acid content 30 to 40 percent; phthalic anhydride 50 to 40 percent; acid value 5 to 25.

Plasticizers

A characteristic of cellulose esters is considerable swelling on addition of solvent before actual dispersion or solution. When a film of the ester solution dries, the process is reversed, resulting in a contraction of the film and a tendency to pull away from the surface. The function of a plasticizer is to correct this tendency and to confer elasticity on the film without destroying the hardness. The ratio of plasticizer to cellulose ester is important and generally is higher for the low-viscosity grades. If a proportion of ethyl cellulose is used in the lacquer, the amount of plasticizer can be reduced.

The ideal plasticizer for high-quality lacquers would be freely soluble in the solvents used, completely compatible with the ester and resin, non-volatile, and chemically inert. With such a material the film would retain its elasticity indefinitely. In practice it is difficult to obtain the desired properties with one plasticizer, and blends are generally used. Even so, the ideal requirements are never completely met.

A great many plasticizers are available and the following are in common use.

Dibutyl phthalate	Tricresyl phosphate
Dioctyl phthalate	Triphenyl phosphate
Dimethylglycol phthalate	Trichlorethyl phosphate
Tributyl citrate	Castor oil

Solvents (cf. Solubility Parameter, p. 147)

The volatile portion of a lacquer usually consists of a blend of liquids, and these are conveniently divided into three groups, based on solvent power.

True solvents form the major portion of the volatile constituents.

Latent solvents are not solvents for the principal film-formers, but increase the solvent power of the other volatile ingredients.

Diluents are not solvents for the principal film-formers but are added to reduce cost. They may also improve the solvent power of the mixture for the resin and plasticizer.

The range of true solvents is not extensive, and typical solvents for cellulose nitrate are esters, ketones and glycol ethers. Examples of the first two classes are:

Esters	*Ketones*
ethyl acetate	acetone
isopropyl acetate	methyl ethyl ketone
n-butyl acetate	methyl isobutyl ketone
ethyl lactate	methyl cyclohexanone

Evaporation rate

The rate of evaporation of a solvent is one of its most important properties and is usually evaluated against ether. A solvent which evaporates too rapidly can cause severe chilling of the film to a temperature below the dew point; and moisture deposition on the soft film can lead to blushing. Also, if the true solvent leaves the film before the latent solvent or diluent, precipitation of the film-former can take place.

Viscosity

The viscosity of a solution of cellulose nitrate depends on the concentration and grade of ester used, the composition of the solvent mixture, and the temperature. The viscosity falls with increasing temperature as is usual. The effect of solvents is not as simple. With simple solvents, viscosity of the solution is lower with low molecular weight solvents than with those of high molecular weight. A mixture of solvents generally gives a lower viscosity than the same proportion of a single solvent. Thus a mixture of alcohol and butyl acetate gives a solution of much lower viscosity than butyl acetate alone. The alcohol here is acting as a *latent solvent*. Alcohols alone are not solvents for cellulose nitrate but they possess a latent solvent power when mixed with other solvents. In practice an alcohol is usually present as the damping agent for cellulose nitrate and its presence is therefore an advantage.

Diluents

In order to keep the cost of a lacquer as low as possible, only sufficient solvent is used to dissolve the cellulose nitrate, the final consistency being attained by dilution with an aromatic or petroleum hydrocarbon. These are not actual solvents but are compatible with the mixture. The proportion of diluent used will depend on its compatibility with the active solvent and this is determined by the dilution ratio. This is given by the expression

$$\text{Dilution ratio} = \frac{\text{Volume of diluent which will just precipitate the dissolved cellulose nitrate}}{\text{Volume of solvent used}}$$

The determination is carried out by adding the diluent to a solution of cellulose nitrate in the specified solvent until the cellulose nitrate commences to precipitate and will not redissolve in the mixture. The value depends on the type of cellulose nitrate and the concentration. This is generally 10 percent. The following dilution ratios are quoted by Durrans [2].

Solvent	Toluene	Xylene	Petroleum	Butanol
Acetone – – – – –	4·5	3·9	0·6	7·0
Amyl acetate (n) – – –	2·2	2·2	1·4	7·3
Butyl acetate (n) – – –	2·7	2·4	1·4	7·3
Butyl phthalate – – –	2·8	2·7	1·7	8·0
Butyl tartrate – – –	10·6	7·7	1·4	15·0
Di-acetone alcohol – –	3·1	2·9	0·5	7·8
Ethyl lactate – – –	5·6	4·8	0·7	10·2
Glycol mono-ethyl ether – –	5·0	4·7	1·0	6·9
Methyl ethyl ketone – –	4·5	3·3		
Tricresyl phosphate – –	3·3	4·2	0·7	4·9

Pigmented lacquers

The dispersion of pigments in cellulose nitrate lacquers follows broadly the same lines as for gloss paints. The pigment content is generally lower, and the lower viscosities of the media can result in troublesome settlement of heavy pigments. The desired properties of the pigment are maximum opacity, freedom from any tendency to bleed in lacquer solvents, resistance to ultra-violet light and freedom from toxicity (this applies to spray paints and any finishes for children's playthings).

The majority of pigments used in gloss paints can also be used in cellulose nitrate lacquers, but care should be exercised with the following, which are either unsuitable or show the defects mentioned:

Zinc oxide — Reactive. Poor opacity but gives good gloss and colour. Can be used for interior work.

Ultramarine Blue — Poor strength and opacity.

Phthalocyanine Blue — Used mainly for tinting. When used alone shows tendency to give bronzy films of poor gloss.

Hansa Yellows   Tendency to bleed.
Ferrite Yellows   Poor gloss. Used for tinting.
Helio Red   Tendency to bleed.
Maroons   Many show poor exterior durability.
Lead Chromes ⎫
Chrome Greens ⎭   Toxic.

Dispersion of the pigment can be effected in a ball or pebble mill, or the pigment can be dispersed in plasticizer alone on a triple roll mill after pre-mixing. A slow evaporating solvent can be used to ease the material over the rolls, but the process is slow.

Chip method. This method is now widely used and consists in dispersing the pigment in a plastic cellulose dough in a heavy mixer. Dispersion of the pigment is effected by high shear forces resulting from high power input and high consistency. The dispersion process is carried out in a Bridge–Banbury mixer which incorporates a pair of bladed rotors turning in a closed chamber. The latter is jacketed for heating or cooling. The material to be processed is forced between the blades by hydraulic ram, and dispersion is effected in a short time.

In the chip process, the damp cellulose nitrate, pigment, and solvent (often a solvent plasticizer) are charged into the mill. When dispersion is complete the mass is passed through sheeting rolls and reduced to a thickness of about 2 mm ($\frac{1}{16}$ inch). The sheets are then broken up into chips.

Several companies now specialize in the supply of chips to the industry, and their use makes the manufacture of pigmented lacquers a relatively simple operation.

CELLULOSE ACETATE

Cellulose acetate was first introduced in 1865 by Schützenberger who prepared it by treating cellulose with acetic acid and acetic anhydride. The formation of cellulose triacetate can be represented by the equation

$$C_6H_7O_2(OH)_3 + 3CH_3COOH \rightarrow C_6H_7O_2(OCOCH_3)_3 + 3H_2O$$

The triacetate is of little interest on account of limited solubility, but by controlled hydrolysis part of the acetate groups can be split off, yielding a product soluble in a number of solvents and useful for lacquer manufacture. It is usual to assess cellulose acetate by the acetyl value, which is the weight (grams) of acetic acid liberated on hydrolysis of 100 grams of ester.

Manufacture of cellulose acetate

Two methods are in use at the present time.

Dreyfus method. This process involves a pretreatment of purified cotton linters or wood pulp with acetic acid containing a small amount of sulphuric acid. This serves to break down the cellulose structure and to increase the reactivity. The product is then added to a mixture of glacial acetic acid and acetic anhydride, the temperature being kept at 4° to 5°C (39·2° to 41·0°F). When the reaction is complete, a clear viscous solution of triacetate is obtained which is then subjected to hydrolysis or "ripening". Water is added gradually and the whole then heated to 60° to 70°C (140° to 158°F) to effect the hydrolysis. For lacquer manufacture an acetyl value of 54 to 55 is considered suitable.

Methylene chloride process. Dry cellulose is treated first with acetic anhydride and then with a mixture of acetic anhydride, methylene chloride and sulphuric acid. Rigid control of conditions, especially temperature, is essential. When the reaction is complete, some water is added and the mixture hydrolysed to the desired degree. The sulphuric acid is then neutralized by sodium acetate, and the methylene chloride recovered by distillation. The cellulose acetate is precipitated from the acetic acid solution by addition of water, washed well, and dried.

Properties

The cellulose acetates are white granular materials with specific gravities of about 1·36. They are thermoplastic and possess high but indefinite melting points. In the molten condition they decompose rapidly.

The general properties are determined largely by the degree of acetylation and the viscosity. They are soluble in a limited range of solvents, the most important of which are methyl acetate, ethyl acetate, acetone, methylethyl ketone, methylcellosolve acetate, carbitol acetate and diacetone alcohol. In addition they dissolve in mixtures of ethyl alcohol (10 parts) with methylene chloride or ethylene chloride (90 parts).

Films of cellulose acetate are very much less flammable than those of cellulose nitrate but have inferior water resistance. They are very resistant to degradation by sunlight and are very resistant to petrol, oils and greases. They are not resistant to alkali. For best results, plasticizers such as triacetin or phthalates are incorporated in quantities up to about 30 percent.

Cellulose acetates are used in lacquers for wood, metal, paper and leather; in strippable coatings, and in paint strippers.

Cellulose acetobutyrate

This material is in some ways very similar to cellulose acetate but possesses several advantages. It is more resistant to water, is soluble in a wider range of solvents, and is compatible with a greater number of plasticizers and resins.

Cellulose acetobutyrate can be prepared in methylene chloride solution using either a mixture of acetic and butyric anhydrides, or a mixture of acetobutyric anhydride, acetic anhydride and a small quantity of sulphuric acid.

The ratio of butyric to acetic anhydride can be varied to give a range of products in which flexibility, solubility and compatibility increase with increasing butyric content. They are used in lacquers for wood, wire, paper and plastics.

Safety Regulations

The highly flammable nature of cellulose nitrate and of the solvents used in cellulose lacquers has resulted in the issue of statutory regulations governing the manufacture, storage and use of these lacquers.

Cellulose nitrate is classed as explosive if the nitrogen content is greater than 12·3 percent. The type used in lacquers contains between 10·5 and 12·3 percent nitrogen and is supplied damped with at least 25 percent of butanol, isopropanol or ethanol.

Cellulose lacquers and thinners containing hydrocarbon and other solvents are controlled by the Highly Flammable Liquids and Liquefied Petroleum Gases Regulations, 1972. The regulations control the storage, use and labelling of highly flammable liquids which are defined as substances which give off a flammable vapour at a temperature less than 32°C (89·6°F) (Abel) and, in addition, support combustion. The method of test is described in Schedule 2 of the Regulations. These Regulations extend the classification of highly flammable liquids from a flash-point limit of 22·77°C to 32·22°C.

Regulations also exist controlling the labelling of containers and the conveyance by road of flammable liquids.

CELLULOSE ETHERS

Ethyl cellulose

If purified cellulose is treated with concentrated sodium hydroxide solution an *alkali cellulose* is produced by reaction with the hydroxyl groups:

$$C_6H_7O_2(OH)_3 + 3NaOH \rightarrow C_6H_7O_2(ONa)_3 + 3H_2O$$

The sodium compound can be reacted with an alkyl chloride such as ethyl chloride under appropriate conditions to give a tri-ether:

$$C_6H_7O_2(ONa)_3 + 3C_2H_5Cl \rightarrow C_6H_7O_2(OC_2H_5)_3 + 3NaCl$$

The method of manufacture is based on these reactions, the conditions being rigidly controlled to ensure the correct degree of ethylation. The tri-ether is of no interest as it has poor solubility and compatibility and is not thermoplastic. The type found most suitable contains between 2 and 2·5 ethoxyl groups per cellulose "unit" which is equivalent to 42 to 48 percent ethoxyl content.

When the desired degree of ethylation has been reached, the excess of ethyl chloride is removed and the ethyl cellulose precipitated by addition of water. The material is washed free from chloride and dried.

Properties

Ethyl cellulose is a white granular powder with specific gravity 1·14. It softens at about 140°C (284°F) and melts at between 200° and 220°C (392° and 428°F). The degree of flammability is very low. It is resistant to dilute acids and to alkalis.

Films of ethyl cellulose are tough and flexible, retaining the latter property over a wide range of temperature. The films are transparent and transmit ultraviolet light without discoloration. The water permeability is superior to that of cellulose acetate but inferior to that of cellulose nitrate.

Ethyl cellulose has the widest range of solubility of all the cellulose derivatives. It is soluble in alcohols, ketones, esters, aromatic and chlorinated hydrocarbons. The range of solubility is controlled by the degree of ethylation, reaching a maximum when about 2·5 of the three available hydroxyl groups are combined. It is also compatible with many drying oils, varnishes, and synthetic resins.

Uses

Ethyl cellulose is incorporated into oleoresinous varnishes to improve drying characteristics, hardness and water resistance. It is also used in flow-back lacquers, cable lacquers and coatings for foil and paper.

Water-soluble cellulose derivatives

These materials absorb water and slowly form gels or "solutions" of high viscosity at low concentrations. With the exception of sodium carboxymethyl cellulose (SCMC) they are ethers which are tasteless and non-toxic. In addition to their extensive use as thickeners and protective colloids in latex emulsion paints they are used in foodstuffs. The most widely used are the following:

Methyl cellulose. A methyl ether of cellulose which is obtained by methylating alkali cellulose with dimethyl sulphate or methyl chloride. The degree of etherification can vary and this controls the properties. When the number of methoxy groups per anhydroglucose unit lies between 1·3 and 2·4 the product is soluble in water, but with increasing methoxy groups it becomes soluble in organic solvents [3].

Aqueous solutions of methyl cellulose are non-ionic and have surface tensions lower than that of water; consequently they form foams on vigorous agitation. The solutions form firm gels which coagulate reversibly on heating and require protection against bacterial attack.

When used as a thickener and protective colloid in emulsion paints the films show good moisture resistance and sizing properties, but in brushability and storage stability they are inferior to hydroxyethyl cellulose and SCMC.

Hydroxyethyl cellulose. An ether in which the degree of etherification is approximately 8·5 hydroxyethyl groups per 10 anhydroglucose units. It produces firm gels and is soluble in both hot and cold water. The solutions are non-foaming and show excellent tolerance to electrolytes.

In latex emulsion paints it gives good scrub resistance, brushability, flow out and sizing properties on porous surfaces. In common with other cellulose ethers it requires protection against bacteria.

Sodium carboxymethyl cellulose (SCMC). This is prepared by reacting alkali cellulose with sodium monochloracetate. The degree of substitution is of the order of 7 carboxymethyl groups per 10 anhydroglucose units. The solutions in water are soft gels and exhibit pseudoplastic flow. They are anionic with a pH of 7·0 and are stable between pH 4 and pH 10.

SCMC gives precipitates with trivalent and some divalent metals, and when used as a protective colloid in emulsion paints it is usual to add a sequestering agent such as Calgon [4]. The emulsion paints show excellent brushing and storage properties but are inferior in other respects to those based on methyl and hydroxyethyl cellulose. Like the latter, SCMC solutions require protection against bacterial attack.

The choice of cellulose thickener for a specific purpose will be influenced by the relative importance of the desired properties. Each product has strong features as well as weaknesses, and a relative assessment is necessary before deciding on the type to be used.

References to Chapter 15

[1]   OCCA Paint Technology Manuals, Part 1. Chapman & Hall, London, 1961.
[2]   *Durrans' Solvents*, 8th edn, revised by E. H. Davies. Chapman & Hall, London, 1971.
[3]   "Celacol" water-soluble cellulose ethers. British Celanese Ltd.
[4]   Albright & Wilson Ltd.

16   Varnishes and aqueous media

VARNISHES

The term varnish is applied to a class of clear liquids of 2 to 3 poise viscosity which, when exposed to the air in thin coatings, dry to tough, glossy films. Traditional varnishes are of two types, known as "oil" and "spirit" varnishes.

OIL VARNISHES

Oil varnishes are combinations of drying oils and resins which are thinned to application viscosity with volatile solvent. Driers are incorporated to catalyse the drying process, and this will be controlled largely by the amount of oil present or the ratio of oil to resin. This ratio, by weight, is known as the *oil length* and can be varied considerably. Varnishes are described as *long*, *medium*, or *short* oil length, and these have roughly the following oil/resin ratios:

Long oil	3/1 to $4\frac{1}{2}$/1
Medium oil	2/1 to 3/1
Short oil	1/3 to 2/1

In the U.S.A. oil lengths are expressed in gallons (U.S.) of oil per 100 lb resin, so that a "10 gallon" oil length varnish would consist of 10 U.S. gallons of oil to 100 lb resin. (Note: 1 U.S. gallon = 0·833 Imperial gallon = 3·78 litres.)

Some of the short-oil varnishes retain a surface tack for some time and have been used as adhesives for gold leaf. Such varnishes are, in consequence, known as *goldsizes*.

The main use of oil varnishes has been the finishing of wood. The wood was either stained or grained before varnishing, although in some cases, where the wood presented a naturally attractive appearance, the varnish was applied without the preliminary staining. As a result of changing fashions in the use of wood and the growth of mass-produced furniture and wooden articles generally, oil varnishes for interior work have been largely replaced by lacquer type finishes, notably cellulose nitrate, polyurethanes and polyesters.

The term "mixing varnish" has long been applied to oil varnishes which are used as paint media. Before the widespread use of alkyds, these mixing varnishes were the standard media for most gloss paints and undercoats, but

263

today their uses are limited to certain metallic and maintenance paints.

Although oil varnishes are now manufactured on a reduced scale and the use of fossil resins is practically obsolete, a short account of these materials will, it is hoped, be of interest.

Oil varnishes contain the following four major ingredients:

Resins, which provide the hardness and gloss of the film. Both these properties increase with increasing resin content, but the durability decreases. Natural resins, especially the fossil copals, were formerly used extensively (*see below*), but these resins are no longer economically viable. Tung oil varnishes for special purposes were often made with ester gum or limed rosin. Resins used in modern varnishes include pure and modified phenolics, maleics, coumarone, and hydrocarbon types.

Drying oils contribute the elasticity, toughness, and general durability. These properties improve with increasing oil content, but the effect is controlled by the type of oil. The commonly used oils are linseed (as VLO, stand oil, or enamel oil) and tung oil. Varnishes often contain both oils.

Driers catalyse the conversion (by oxidation and polymerization) of the oil to the solid state. They are usually added as solutions of naphthenates or octoates of lead, cobalt, and sometimes manganese. The desire to eliminate lead from all types of coatings has stimulated the search for other polyvalent metals which are non-toxic. Mixtures of zirconium, cobalt and calcium as naphthenates or octoates appear to be promising alternatives to lead/cobalt mixtures.

Thinners are added for the sole purpose of reducing the oil/resin complex to a suitable consistency for application. They should evaporate completely from the applied film. White spirit is the usual thinner for long-oil varnishes, but the shorter types, and especially some tung oil varnishes, require a proportion of stronger solvent such as xylene, naphtha, or aromatic petroleum solvent ("Shellsol", "Solvesso" or "Aromasol") [1].

Varnishes from fossil resins

The industrial production of durable oil varnishes was originally based on the combination of linseed oil with African hard fossils such as Animi and Sierra Leone copals. In their natural condition these resins are not soluble in drying oils but can be made soluble by heating for some time at a temperature just above the fusion point. This results in a partial depolymerization during which about 20 to 30 percent of the weight of the resin is lost in volatile decomposition products. This process is known as *gum running*.

Congo copal has been used on a large scale by varnish manufacturers but has also been run in bulk, allowed to cool, and marketed as *fused copal*, or esterified with glycerol and marketed as *copal ester*.

The varnish-making operation consists of the following successive steps:

(a) running the resin, (b) incorporating the run resin with the linseed oil, (c) cooling, dilution with thinner and addition of driers, (d) clarification and maturing. Clarification is necessary since the natural resins always contain a certain amount of extraneous matter.

Properties and uses of copal varnishes

The drying times of these varnishes vary with the oil length, the long-oil types taking from 16 to 20 hours and the short-oils from 1 to 2 hours.

The term "copal varnish" has been used in the decorating trade for the best-quality varnish. Long-oil types are used for exterior work and medium-oil types for interior decoration. The long-oil types were formerly used as carriage varnishes, and for many years were the standard type of medium for hard-gloss paints. Other types of varnish made on copal resins are the short-oil types known as "hard church oak", "flatting varnish", and the goldsizes.

VARNISHES FROM MODIFIED PHENOLIC RESINS

Modified phenolic resins are soluble in drying oils at comparatively low temperatures, so the varnish-making process is relatively simple. Again, the wide range of rosin-modified phenolic resins available enables varnishes to be designed to meet a wide spectrum of requirements.

Some of the softer types of modified phenolic resin are soluble in white spirit in the cold and can be made into varnishes by solution in this solvent followed by addition of bodied oil and driers. This process is known as *cold-cutting*.

With the harder types, the resin and part of the linseed oil are heated to 250° to 260°C (482° to 500°F) until a drop forms a clear bead on cooling on glass. A second portion of oil is added and the test repeated. The process is continued until all the oil has been added and the varnish gives a clear solution in white spirit. It is then cooled and thinned, and driers are added in the normal way.

The softer types of modified phenolic resin can be treated in a similar way, but the oil and resin can be blended in fewer steps.

These varnishes are very similar in their general properties to those made from Congo copal but they are more easily made.

Tung oil is often incorporated into these varnishes along with the linseed oil and is best used in the form of enamel oil. These oils contain 20 to 25 percent tung oil co-polymerized with linseed oil. Processing is similar to that for linseed oil, but the varnishes have appreciably greater water resistance.

TUNG OIL VARNISHES

Tung (or wood) oil can be used in varnish-making, either in the form of a stand oil (these are available in a range of viscosities) or as the cheaper raw tung oil. The stand oil can be blended with the softer modified phenolic resins or with coumarone resins at 200° to 220°C (392° to 428°F), but harder resins require a higher temperature, with its attendant risk of gelation. To reduce this risk the resin is best cooked with linseed oil and oiled out with the tung-oil stand oil. It is then cooled rapidly and thinned.

It is more economical to use raw tung oil and to effect the bodying during the processing, but this method requires close control and experience. The raw tung oil is mixed with half the resin and heated rapidly in an open pot to 280°C (536°F). This temperature is necessary to ensure that the tung oil is *gas proof*, i.e. that the film on drying will be unaffected by gases from combustion. It must be borne in mind that tung oil will gel in 12 to 20 minutes at 280° to 285°C (536° to 545°F), so all operations must be carried out smoothly and without delay.

When the temperature has reached 280°C the pot is moved away from the fire, and after about 2 minutes the remainder of the resin, in small pieces, is added rapidly, with stirring. The temperature of the mixture falls to 250° to 260°C (482° to 500°F) at which temperature bodying is completed. It is then cooled rapidly and thinned.

Many tung-oil varnishes contain a proportion of linseed oil which is often useful as a *chill-back* if the tung-oil bodying tends to get out of control.

Tung oil–pure (100 percent) phenolic varnishes

The two types of pure phenolic resin (Chapter 13) behave differently towards tung oil, the resole type being reactive and the novolac type being non-reactive but soluble in the oil.

Reactive resole resins. The manufacture of varnishes from these resins is effected by heating the tung oil to 170° to 180°C (338° to 356°F), at which point the resin is added and the whole stirred until the resin is dissolved. The temperature is then raised to 230°C (446°F) when reaction occurs accompanied by foaming. The intensity of foaming varies from one resin to another, and care must be taken to regulate the temperature so that the reaction proceeds slowly, otherwise there is a risk of the pot boiling over. The melt is held for about 25 to 30 minutes, the end-point being indicated by lessening or cessation of foaming. It is then cooled and thinned.

It is of interest that with reactive resins a gas-proof varnish is obtained at the relatively low temperature of 230°C (446°F).

Non-reactive novolac resins. These resins dissolve in tung oil but no reaction occurs between them, so bodying involves polymerization of the oil only. The varnishes are made in the normal way for tung-oil media, but the thinners must contain a high percentage of aromatic hydrocarbons as the resins have limited solubility in white spirit.

Properties and uses of tung oil–phenolic varnishes

These varnishes are all characterized by higher water resistance than linseed oil types, the highest water resistance and the quickest dry being shown by the 100 percent phenolic types. Of these, varnishes from the reactive types are the more popular. They are used as spar varnishes and as media for certain types of maintenance paints, for example, those pigmented with micaceous iron oxide and used on structural steelwork.

Tung oil–coumarone varnishes

These are usually made at about 2/1 oil length and are conveniently manufactured from bodied tung oil. The oil and resin are heated to 220°–230°C (430°–450°F) until a uniform melt is obtained, cooled rapidly, and thinned with white spirit. The alkali resistance of these varnishes is of a high order and they are used in primers for alkaline surfaces such as plasters, cement render, and asbestos board. The low acid value of these media renders them very suitable for use in metallic paints.

PHYSICAL PROPERTIES OF OIL VARNISHES AND FILMS

Colour. This is influenced by the colour of the oils and resins used and also by the care taken in manufacture. Excessive heating is a frequent cause of dark colour. The measurement of the colour of varnishes is carried out by methods similar to those used for oils (Chapter 10).

Viscosity. Varnishes are generally applied by brush and the brushing properties are determined largely by the viscosity. If this is too high, brush-drag will be experienced and the varnish will be difficult to spread; if too low, the varnish will form runs on vertical surfaces and the film will be excessively thin. It is found that a viscosity of 2 to 3 poises at 25°C (77°F) is generally satisfactory for a brushing varnish. Viscosity measurements are carried out by the same methods as those used for oils (Chapter 10).

Drying time. The drying time of a varnish is dependent on the composition but, in addition, it is influenced by film thickness, temperature and humidity of the air, nature of the substrate, and the amount of light reaching the surface. The methods for the determination of the drying times of varnishes are the same as those used for paints and are described in BS 3900, C2 and C3 [2].

Flexibility. One of the functions of the oil in a varnish is to contribute flexibility, a property which enables a varnish film to accommodate the dimensional changes in the substrate without cracking. This is particularly important on exterior woodwork when expansion and contraction imposes great stresses on the varnish. The *cold-check* test is often applied to varnishes for woodwork. A wooden panel coated with the varnish in question is subjected to a number of cycles consisting of exposure to warm (60°C, 140°F) and cold (−10°C, 23·3°F) conditions. Varnishes which lack flexibility will be found to crack after a few cycles.

The general methods for the determination of the flexibility of varnishes are identical with those used for paints [3].

Hardness. This property, like many others, cannot be assessed in isolation. The property which is important from a practical viewpoint is the resistance of the film to mechanical injury, and this depends on a combination of hardness and toughness. The methods for measuring these are identical with those used for paint [4].

Moisture resistance. For a given type of varnish, the long-oil members are more permeable than the short-oil, owing to the fact that oils are more permeable than resins. Tung oil varnishes are much less permeable than the linseed oil types.

The rate and degree of moisture penetration depends on a number of factors, two of the most important being the presence of polar (hydroxyl, carbonyl, etc.) groups which initiate absorption, and the presence of water-soluble materials such as scission products in the film. These cause osmosis to occur and it is possible for considerable quantities of moisture to be taken into the film. This can result in swelling of the film, and if this is appreciable, blistering can result.

Water absorption comparisons can be made by drying films on glass strips and immersing these in distilled water at constant temperature for definite periods of time. The strips are removed, dried rapidly with blotting paper, and weighed. Glass strips are used so that any milkiness or cloudiness can be observed.

Non-volatile content. This can be determined by weighing 1 to 2 grams of varnish into a circular flat-bottomed metal dish 75 mm in diameter and heating for 3 hours at 105° to 110°C (221° to 230°F). It is then cooled and weighed. Difficulties are sometimes encountered with this method due either to skinning of the sample or to retention of solvents by certain types of resin. The method is useful for checking and correcting solvent losses which take place during the thinning of varnishes.

Flash points and **specific gravities** of varnishes are determined by the methods used for solvents and oils respectively.

SPIRIT VARNISHES

These consist of solutions of resins in volatile solvents, and dry entirely by evaporation. The resin film remains soluble in the parent solvent and is sometimes termed a *non-convertible* coating.

Shellac is the most widely used resin in this class and is dissolved in Industrial Methylated Spirit (IMS) by placing the resin and solvent in a wooden barrel churn. This is rotated until the resin is in solution. A number of spirit varnishes and "polishes" are made from shellac alone or in admixture with other resins. The two members in current use are French polish and knotting.

French polish consists of a 25–30 percent solution of orange shellac in IMS. It is used for polishing good-quality furniture, especially antiques.

Knotting, a 35–40 percent solution of orange shellac in IMS, is used, as the name suggests, for sealing knots in new timber. The purpose is the prevention of exudation of resinous materials which would soften and discolour paint films.

Shellac solutions in ethanol are highly acidic and can cause rapid corrosion leading to discoloration if packed in tinplate containers. The corrosion can be prevented by addition of a small quantity of ethylaniline phosphate.

Polyvinyl acetate is soluble in IMS to give water-white solutions of zero acid value. Such solutions require a plasticizer but yield films of limited weather resistance. They are used as coatings on polished metals and also as media for bronze and aluminium paints for interior use.

AQUEOUS MEDIA

The use of water as a paint solvent has always been attractive for reasons of low cost and absence of fire risk. The increasing concern with environmental pollution has given further impetus to the development of water-borne systems.

The aqueous media in use today can be divided into two main classes: (a) polymer emulsions, which are used chiefly in building paints, and (b) water-soluble or water-thinnable binders used mainly for industrial paints, especially in metal primers applied by electrodeposition. An obviously attractive application of water-soluble binders is in air-drying decorative finishes and undercoats; considerable development work is being carried out to this end.

POLYMER EMULSIONS

Nature of emulsions

An emulsion is a two-phase system consisting of a liquid, in the form of very fine droplets (the dispersed phase), suspended or dispersed in a second liquid (the continuous phase) in which it is insoluble. The commonest form of emulsion contains water as the continuous phase; polymer emulsions are of this type. When the dispersed phase is a solid the system is termed a "dispersion", but although the polymer is usually more solid than liquid, the term "emulsion" is always used for the polymer systems. (Emulsions exist in which water is dispersed in an organic continuous phase, but these are rarely encountered in surface coatings.)

Formation of emulsions

Suppose that it is desired to produce an emulsion of linseed oil in water. Rapid stirring of the two liquids together will produce a suspension of oil droplets, but when stirring ceases, the two liquids will separate into clearly defined layers (Fig. 16.1). If, however, the water contains a small quantity of a soap or a surface-active agent (or "surfactant") the oil droplets remain

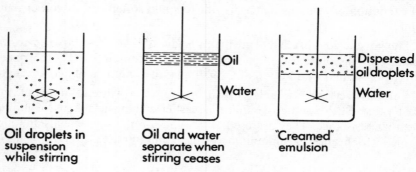

Oil droplets in suspension while stirring   Oil and water separate when stirring ceases   "Creamed" emulsion

Fig. 16.1

dispersed in the water and an emulsion is formed. When this emulsion is allowed to stand for some time, the dispersed oil droplets will tend to rise (the specific gravity of linseed oil being lower than that of water). This is known as "creaming" of emulsions, by analogy with the separation of cream from milk. The "cream" is readily redispersed on stirring, but the formation of "cream" can be prevented by increasing the viscosity of the water phase. Animal glues, alginates, alkali caseinate or cellulose ethers can be used for this purpose. Such thickening agents are also described as "emulsion stabilizers" or "protective colloids".

The oil-bound or washable distempers, once extensively used but now virtually obsolete, were made of an emulsion consisting of bodied linseed oil (or linseed oil and resin) dispersed in a solution of animal glue or alkali caseinate.

Surfactants

These are compounds which, when dissolved in a liquid, reduce the surface tension of that liquid or the interfacial tension between two immiscible liquids or between the liquid and a solid. Their effectiveness is related to their chemical constitution and a very large number of these compounds are known. When a surfactant is used to produce an emulsion, as in the example given above, it is classed as an "emulsifying agent", but, as will appear below, only certain types of surfactant can be used for this purpose.

Surfactants are classified into four main groups designated anionic, non-ionic, cationic, and ampholytic. The anionics are the more commonly used as emulsifying agents. The other classes are used in pigment wetting and dispersion and for special purposes.

Anionic surfactants

The most common types are the alkali metal salts of straight chain carboxylic or sulphonic acids of 11 to 17 carbon atoms. Sodium laurate is a typical example. When dissolved in water it ionizes:

$$CH_3(CH_2)_{10}COONa \rightleftharpoons CH_3(CH_2)_{10}COO^- + Na^+$$

The surface activity is due to the polar nature of the anion which consists of the hydrophilic (water-attracting) COO^- group and the hydrophobic or lipophilic (water-repelling or oil-attracting) hydrocarbon chain. Except at very low concentrations the anions form micelles (Fig. 16.2, where the anions are represented as ——o) in which the hydrocarbon ends are directed inwards and the carboxyl groups outwards into the water:

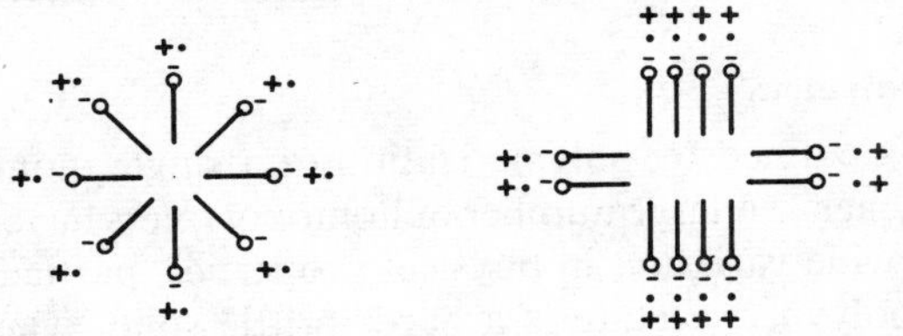

Fig. 16.2

When oil droplets are introduced (as in the introductory example) the hydrocarbon ends are attracted to the oil and the carboxyl groups project into the water (Fig. 16.3):

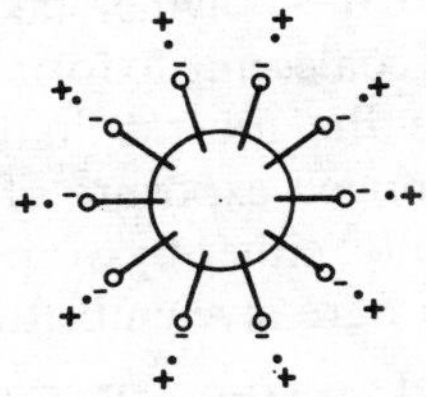

Fig. 16.3

The oil droplets thereby acquire an overall negative charge and are mutually repellent. Coalescence of the droplets is prevented and an anionic emulsion results. The anions tend to concentrate at the surface of the solution and lower the surface tension by an amount proportional to the concentration. In a similar way they are adsorbed at the interface between the solution and a solid and so reduce the contact angle. Wetting of the surface is thereby facilitated.

Other types of compounds used as anionic surfactants are water-soluble salts of amines and fatty acids, e.g. triethanolamine oleate; sulphated higher alcohols, e.g. sodium lauryl sulphate; sulphonated compounds, e.g. sulphonated castor oil (Turkey Red Oil), and aromatic petroleum sulphonates.

Formation of polymer emulsions [5]

When a liquid monomer, such as vinyl acetate, is added to an aqueous solution of an anionic surfactant under rapid stirring, droplets of various sizes are produced and these become coated with the anions. The smaller droplets become enclosed in micelles and are rapidly "solubilized". Addition of the polymerization catalyst results in rapid and vigorous reaction in which the solubilized monomer is converted to the polymer (polyvinyl acetate). The larger droplets are solubilized more slowly, but ultimately the whole of the monomer is converted to polymer. The polymer particles are coated with anions, so acquiring a negative charge. The final system is an anionic emulsion. Emulsions produced in this way contain 45 to 55 percent of polymer by mass.

The polymerization reactions have been described in Chapter 13.

Homo- and co-polymers

The product obtained by polymerization of a single monomer is known as a "homopolymer". A large number of homopolymers have been produced and they show wide variation in physical properties, particularly hardness. For example, polyvinyl acetate is a hard, brittle solid, whereas polybutyl acrylate is soft and flexible in consequence of the different groups in the chains. Polyvinyl acetate therefore does not form continuous films when the emulsion is applied to a surface and allowed to dry. If, however, a suitable plasticizer (up to 20 percent by mass on the polymer) is stirred into the emulsion, migration into the polymer takes place, resulting in softer particles which are capable of coalescing to form a film.

Ester-type plasticizers are the most suitable but, unfortunately, they possess drawbacks. On prolonged exposure of the plasticized film, loss of plasticizer takes place gradually, either by evaporation or by migration into the substrate. The result is progressive embrittlement, leading to failure.

Co-polymers. If vinyl acetate is co-polymerized with a second monomer, which itself gives a soft, flexible polymer, then by adjustment of the relative quantities, products can be produced with any desired degree of hardness and which are capable of yielding continuous films without addition of plasticizer. The proportion and type of second monomer can be varied, and the types used include ethylene and other olefines, vinyl chloride, VeoVa [6] (a vinyl ester of a branched chain carboxylic acid), and acrylic esters. In addition, terpolymers with interesting properties are produced by co-polymerizing vinyl acetate with both ethylene and vinyl chloride.

Vinyl co-polymers are the standard media for the majority of emulsion paints for interior use. Other vinyl co-polymers with more specific applications include vinyl chloride/VeoVa, styrene/acrylic and vinyl toluene/acrylic.

Properties and uses of vinyl emulsions

Plasticizer loss leading to embrittlement has resulted in homopolymer types being replaced by co-polymers. The latter are also more resistant to repeated freezing and thawing. Co-polymers show greater adhesion, with the result that they can be applied in conditions where the homopolymers were unsuitable, e.g. in kitchens, bathrooms and over old paintwork. They are used on both interior and exterior wall surfaces; on the latter the durability is good but not quite as good as that of acrylic emulsions. Co-polymer emulsion paints are textured to produce imitation stone paints for exterior use.

Other uses are as primers for a number of porous surfaces such as hardboard and various types of building boards.

Acrylic emulsions

These are produced by co-polymerization of esters of acrylic acid and its derivatives. Polymerization of ethyl acrylate, for example, yields an extremely soft polymer of little practical use; methyl methacrylate, on the other hand, yields a hard, brittle product on polymerization. By co-polymerizing the two esters in various proportions very useful polymers are obtained. These yield films of excellent flexibility, adhesion to a variety of surfaces, and durability.

Properties and uses of acrylic emulsions

Acrylic emulsions contain polymers of acrylic esters and have shown a number of outstanding properties. They are very stable, resistant to freezing, and less permeable to moisture than polyvinyl acetate.

When made into paints they show very good chemical resistance, adhesion, flexibility, and durability. An example of their adhesion and flexibility is afforded by the successful use of acrylic emulsions on exterior woodwork over lead-based primers.

Acrylic emulsions are coming into extensive use for alkaline plasters and for external painting over new stucco and cement.

The preparation and properties of acrylic emulsions are described in detail by Allyn. [7]

Film formation from emulsions

This is a physical process which, in its simplest form, consists in loss of water by evaporation, followed by coalescence of the polymer particles to give a continuous adherent film. For each type of co-polymer or homo-polymer/plasticizer mixture there is a minimum temperature below which coalescence of the particles, i.e. film formation, will not take place. This temperature is related to the "Glass transition temperature" (designated Tg) of the polymer. This is the point of transition between a hard solid and a plastic or semi-fluid condition. In the U.K., emulsion paints are often applied at temperatures below the Tg of the polymer, and it is usual, therefore, to include a proportion of "coalescing solvent" in the paint formulation. These solvents can be stirred into the emulsion or paint and dissolve in the polymer particles, resulting in softening and lowering of the Tg. This allows the polymer particles to coalesce, giving a continuous film, and the coalescing solvent evaporates. Solvents used for this purpose include certain glycol esters and ether-esters.

Whilst coalescing solvents are effective aids to film formation, they detract from the resistance of the emulsion to freezing. Soft polymer particles are more prone to coagulate as a result of repeated freeze/thaw cycles than are the harder types.

The significance of Glass transition temperature, its relationship to co-polymer composition, and its effect on film formation and film properties have been studied by Bondy [8].

The following figures illustrate the general physical characteristics of commercial emulsions:

	Vinyl	Acrylic
Type of dispersion	Anionic	Anionic
Solid content	50–60%	45–50%
pH value	4·0–5·5	8·5–9·5
Minimum film-forming temperature	3°–15°C (37·4°–59°F)	0°–12°C (32°–53·6°F)
Particle size	0·3–2 μm	0·1–0·5 μm

WATER-SOLUBLE BINDERS

Early types

Among the earliest binders for pigments used by man were the vegetable gums, some of which have been employed for special purposes down to the present day. Examples are gum tragacanth and gum arabic which are still used in limited quantities. Egg yolk is another material used by medieval painters and which continues to be used as a component of artists' "tempera" medium.

Animal glues, obtained from hides and bones, were in general use until the middle of the present century as binders for "soft" distempers (e.g. ceiling whites) and as protective colloids in oil-bound or washable distempers. These materials are now virtually obsolete.

Casein, a protein obtained from milk, is soluble in dilute alkali and was used as a binder per se in "oil-free" distempers and as a protective colloid in the oil-bound or washable types. Casein possessed one advantage over the animal glues. A solution in dilute ammonia gave a water-insoluble film by loss of the alkali. Alkali caseinate solutions continue to be used on a limited scale as a protective colloid in certain types of emulsion paint.

Alginates, obtained from certain species of seaweeds, form colloidal solutions of high viscosity. They have been used as binders for soft distempers.

Modern types

These can be divided into two classes: (a) alkali silicates, and (b) water-soluble resins. The latter are the more important.

Alkali silicates

Sodium silicate was used as the binder in fire-retarding paints (e.g. for fire curtains and stage scenery) for a number of years. However, the films tended to whiten due to formation of sodium carbonate, and for this purpose sodium silicate was later replaced by potassium silicate. Sodium silicate was also used in paints required to resist moderately high temperatures, but the films were not sufficiently weather-resistant for outside work.

An interesting use of sodium silicate has been in the so-called "inorganic zinc silicate" paints in which zinc dust is mixed with the sodium silicate solution shortly before use. The film consists of zinc silicate containing unreacted zinc and is an effective primer for steel. It is essential, however, that the SiO_2/Na_2O ratio in the sodium silicate be kept within fairly narrow limits.

Lithium silicate was later used to replace sodium silicate in zinc silicate paints, but the use of alkali silicates for this purpose has dropped considerably since the introduction of the prehydrolysed ethyl silicates. A description of the manufacture and uses of alkali silicates has been published by Pass and Meason [9].

Water-soluble resins

A number of synthetic resins can be produced in water-soluble form by incorporating acidic groups, e.g. carboxyl, in the molecular structure. These groups are then neutralized by bases such as ammonia or amines to give products soluble in water.

Water-soluble resins have been used as binders in conventional type stoving enamels for a number of years, but surfaces to be coated must be very clean. The slightest traces of oil or grease can impair wetting of the surface and result in serious cissing or contraction of the film [10].

In 1963, metal primers based on water-soluble resins were introduced in the car industry and applied by electrodeposition. This has probably been the most important application of these resins to date. The process of electrodeposition involves several simultaneous reactions and has been well covered in the literature [11]. The following is a brief outline of the basic reactions.

When a water-soluble resin is dissolved in water, ionization takes place giving a large (resinous) anion and an ammonium (or amino) cation:

$$R.COONH_4 \rightleftharpoons R.COO^- + NH_4^+ \quad (R = resin\ complex)$$

Some resins require a proportion of water-miscible solvent in the solution to give optimum results.

The object to be coated is made an anode, the cathode being either the metal tank surface or a separate electrode. On passing a direct current, deposition of the resin (and associated pigment) takes place on the anode; cure of the coating is effected by stoving. This method — known as "anodic deposition" — has been used widely in spite of certain operational drawbacks such as metal dissolution and staining of the coating.

Recently it has been found possible to carry out cathodic deposition by using water-soluble resins which ionize to give resinous cations. These resins can be produced by various methods, but generally they involve the use of resins containing basic groups which can be neutralized by acid to give water-solubility. In water they ionize according to the scheme:

$$\begin{array}{ccc} R_1 & & R_1 \\ | & & | \\ R_2{-}NHCl & \rightleftharpoons & R_2{-}NH^+ + Cl^- \\ | & & | \\ R_3 & & R_3 \end{array}$$

For example, epoxy resin/amine adducts which can be neutralized with acids to give water-soluble binders for cathodic deposition have been described by van Westremen [12]. A critical comparison of the relative merits of cathodic and anodic deposition has been made by Verdino [13].

The nature of electrodeposition limits the application to single coats, and it is used mainly for primers on steel. The method is in general use in the car industry, and although this consumes a high percentage of the water-soluble resins made, significant amounts are being used in other paint applications.

Types of resin used

Alkyds

These are produced by selecting the appropriate alcohol and acid to give the desired free acid groups in the molecule. A wide range of suitable acids is available, among which are maleic and isophthalic acids and trimellitic anhydride. Glycols are among the alcohols used. A comparison of the performance of water-soluble alkyds with the conventional solvent-based systems has been published by Engelhardt [14].

Burckhardt and Luthardt [15] have examined water-soluble short-oil alkyds, particularly for use by conventional and electrostatic spray. They conclude that their behaviour is determined by the functional groups causing the solubility. Water-soluble methoxylated melamines are often added to cross-link with the alkyd. These water-soluble alkyds are used also as protective colloids for acrylic emulsions.

Epoxy esters

These are among the resins most widely used since the introduction of electrodeposition. Their preparation and uses have been described by van Westremen and Tysall [16]. The use of modified epoxy esters which are hardened by cross-linking with water-soluble melamines or phenolic resins is covered by patents [17]. In a recent publication Bijleveld and Krak [18] describe the solubilization of epoxy esters by maleinization. They also discuss new types of water-soluble alkyds which are under investigation.

Other resins

Other water-soluble resin systems of interest, particularly in electro-deposition, are polyesters, acrylics, amino/formaldehyde and phenol/formaldehyde. These have been described by Hopwood [19] and by Crawford [20].

Appreciable amounts of water-soluble resins are used in paints for application by the coil-coating process. According to Percy and Nouwens [21], the most successful types are the polyester and thermosetting acrylics. Both types can be cross-linked during stoving with water-soluble melamine.

References to Chapter 16

[1]   "Shellsol" (Shell Chemicals); "Solvesso" (Esso Petroleum); "Aromasol" (ICI).
[2]   BS 3900:1965–74. Methods of test for paints. Part C2, Surface drying time; Part C3, Hard drying time. British Standards Institution, 2 Park Lane, London, W.1.
[3]   Ibid., Part E1. Bend test.
[4]   Ibid., Part E2. Scratch test.
[5]   See also FRIIS, N. & NYHAGEN, I., Kinetic study of the emulsion polymerization of vinyl acetate. J. Appl. Polymer Sci., 1973, 17, No. 8, 2311.
[6]   Shell Chemicals Ltd. See also ATEN, W. C. & VEGTER, G. C., New copolymer latices on the basis of VeoVa monomer in paint applications. JOCCA, 1970, 53, 448.
[7]   ALLYN, G., Acrylic ester emulsions and water-soluble resins. In Treatise on Coatings, MYERS, R. R. & LONG, J. S. (eds), Vol. 1, Part 1, Edward Arnold, London, 1967.
[8]   BONDY, C., Binder design and performance in emulsion paints. JOCCA, 1968, 51, 409.
[9]   Pass, A. & MEASON, M. J. F., Alkali silicates in surface coatings. JOCCA, 1965, 48, 897.
[10] HESS' Paint Film Defects, 3rd edn, p. 62. Chapman & Hall, London, 1979.
[11] TAWN, A. R. H. & BERRY, J. R., The electrodeposition of paint—some basic studies. JOCCA. 1965, 48, 790.
      CHANDLER, R. H., Advances in electrophoretic painting. (Published at intervals in series "Bibliographies in Paint Technology".) R. H. Chandler, Braintree, Essex.
      MACHU, W., Handbook of Electropainting Technology. (Translated from German by P. Neufeld.) Electrochemical Publications Ltd, Ayr, Scotland, 1978.
[12] VAN WESTREMEN, W. J., Modern developments in aqueous industrial coatings. JOCCA, 1979, 62, 246.
[13] VERDINO, H., Some examples of electrophoretic coatings for cathodic deposition. JOCCA, 1976, 59, 81.

[14] ENGELHARDT, R. R., Water reducible industrial coatings. *Pig. & Resin Tech.*, 1979, **8,** No. 8, 5.

[15] BURCKHARDT, W. & LUTHARDT, H. J., Water-borne industrial thermosetting systems; physical chemical behaviour of their components and aspects of application technology. *JOCCA*, 1979, **62,** 375.

[16] VAN WESTREMEN, W. J. & TYSALL, L. A., Some aspects of the preparation and testing of recently developed epoxy resin esters. *JOCCA*, 1968, **51,** 108.

[17] American Cyanamid Co., Ger. Pat. 1,669,593 (1966). BASF, Ger. Pat. 1,930,949 (1969); 2,001,232 (1970).

[18] BIJLEVELD, J. & KRAK, H., Some aspects of spray applied water-borne paint binders. *JOCCA*, 1975, **58,** 279.

[19] HOPWOOD, J. J., Water soluble thermosetting organic polymers. *JOCCA*, 1965, **48,** 157.

[20] CRAWFORD, H. R., The design of resins for electrodeposition. *JOCCA*, 1972, **55,** 557.

[21] PERCY, E. J. & NOUWENS, F., Aqueous coil coatings. *JOCCA*, 1979, **62,** 392.

Appendix A

BIBLIOGRAPHY

Colour and Pigments
ABRAHART, E. N., *Dyes & their Intermediates*, 2nd edn. Edward Arnold, London, 1977.
ALLEN, R. L. M., *Colour Chemistry*. Nelson, London, 1971.
CLULOW, F. W., *Colour: its principles and their application*. Fountain Press, 1972.
GRIFFITHS, J., *Colour and Constitution of Organic Molecules*. Academic Press, New York, 1976.
HEATH, F. J., *An Introduction to the CIE System*. Tintometer Ltd, Salisbury, 1967.
HERDAN, G., *Small Particle Statistics*. Butterworth, London, 1960.
ISHIHARA, S., *Tests for Colour Blindness*. Lewis, London, 1959.
JELINEK, Z. K., *Particle Size Analysis*. Wiley, Chichester/New York, 1974.
JUDD, D. B. & WYSZECKI, G., *Color in Business, Science & Industry*, 2nd edn. Wiley, 1963.
PARFITT, G. D. & SING, K. S. W., *Characterisation of Powder Surfaces with Special Reference to Pigments and Fillers*. Academic Press, New York, 1976.
PATTON, T. C. (ed.), *Pigment Handbook*, Vol. 1–3. Wiley, 1973.
STOCKHAM, J. D. & FOCHTMAN, E. G. (eds), *Particle Size Analysis*. Wiley, Chichester/Ann Arbor, 1977.
WRIGHT, W. D., *The Measurement of Colour*, 4th edn. Hilger & Watts, London.

Bitumen, Glues and Gums
ABRAHAM, H., *Asphalts and Allied Substances*, 4 vol. Van Nostrand, 1960–62.
BAGUE, R. H., *The Chemistry and Technology of Gelatin and Glue*. McGraw-Hill, New York.
MANTELL, C. L., *The Water Soluble Gums*. Reinhold, New York, 1947.
SUTERMEISTER, E., *Casein and its Industrial Applications*. Chapman & Hall, London.

Oils, Solvents and Plasticizers
DOOLITTLE, A. K., *Technology of Solvents and Plasticisers*. Wiley, 1954.
DURRANS, T. H., *Solvents*. 8th edn, revised by E. H. Davies. Chapman & Hall, London, 1971.
LAGOWSKI, J. J. (ed.), *The Chemistry of Non-aqueous Solvents*. Academic Press, New York, 1978.
MARSDEN, C. & MANN, S., *Solvents Guide*. Cleaver-Hume, London, 1963.
Oil & Colour Chemists' Association, *Introduction to Paint Technology*. London, 1976.
WADDAMS, A. L., *Chemicals from Petroleum*, 4th edn. John Murray, London, 1978.

Resins
BILLMEYER, F. W., *Textbook of Polymer Chemistry*. Wiley, New York, 1962.
BLACKLEY, D. C., *Emulsion Polymerisation—Theory and Practice*. Science Publishers, 1976.
ELIAS, HANS-GEORG, *Macromolecules*. Vol. 1—*Structure and Properties*. Vol. 2—*Synthesis and Materials*. Wiley, Chichester, 1977.
HILDEBRAND, J. H. & SCOTT, R. L., *The Solubility of Non-Electrolytes*, 3rd edn. Reinhold, 1950.
HOLMES-WALKER, W. A., *Polymer Conversion*. Applied Science, Barking, 1975.
MARK, H. & WHITBY, G. S. (eds), *The collected papers of W. H. Carothers on high polymeric substances*. Interscience, 1950.
SMITH, D. A. (ed.), *Addition Polymers—Formation and Characteristics*. Butterworth, London, 1968.
SOLOMON, D. H., *The Chemistry of Organic Film Formers*. Wiley, New York, 1967.

Appendix B

BRITISH STANDARDS RELATING TO PAINT MATERIALS

Many of the following standards have been referred to in the text. They are obtainable from the British Standards Institution, 2 Park Street, London, W1A 2BS.

Note: This list covers the most commonly used materials but is not exhaustive.

Colour

381C:1964. Colours for specific purposes.
    PD 5824:1966. Supplement No. 1. Table of colorimetric values of colours to BS 381 C.
905:1967. Artificial daylight for the assessment of colour.
4800:1972. Paint colours for building purposes.

Oils

242 (includes 243, 259 & 632):1969. Linseed oils.
391:1962. Tung oil.
4725:1971. Linseed stand oils.

Pigments and Extenders

217:1961. Red lead for paints and jointing compounds.
239 (includes 254, 296, 338 & 637):1967. White pigments for paints. (Contains white lead, zinc oxide, lithopone, antimony oxide, basic lead sulphate.)
282, 389:1963. Lead chromes and zinc chromes for paints.
283:1965. Prussian blues for paints.
284, 285, 286:1952. Black (carbon) pigments for paints.
303:1978. Lead chrome green pigments for paints.
314:1968. Ultramarine pigments.
318:1968. Chromium oxide pigments.
388:1972. Aluminium pigments.
1795:1976. Extenders for paints.
1851:1978. Titanium pigments for paints.
3599/1–5:1963 } Organic pigments for paints.
3599/6–15:1964 }
3699:1964. Calcium plumbate.
3981:1978. Iron oxide pigments for paints.
3982:1966. Zinc dust pigment.
4313:1968. Strontium chromate for paints.
5193:1975. Zinc phosphate pigments for paints.

Resins

1284:1960. Bleached lac.
3279:1960. Seedlac.

Solvents and Plasticizers

135, 458, 805:1977. Specifications for benzene, xylenes and toluenes.
245:1976. Specification for mineral solvents (white spirit and related hydrocarbon solvents) for paints and other purposes.
479:1963. Aromatic naphthas.
507:1966. Ethanol.
508:1966. *n*-Butanol.

551:1965. *n*-Butyl acetate.
573, 574, 1995, 1996, 2535, 2536 & 3647:1973. Plasticizer esters.
1595:1965. Isopropanol.
1940:1968. Butanone (MEK).
1941:1968. 4-methylpentan-2-one (MIBK).
2711:1967. Cyclohexanone.
3591:1963. Industrial methylated spirit.

Test methods

188:1977. Methods for the determination of the viscosity of liquids.
1006:1971. Methods for the determination of the colour fastness of textiles to light and weathering.
2839:1969. Method for the determination of flash point of petroleum products by the Pensky–Martens closed tester.
3406:1961–3. Methods for the determination of the particle size of powders.
3442:1974. Method for the determination of flash point by the Abel apparatus.
3483:1974. Methods for testing pigments for paints.
4359:1969–71. Methods for the determination of the specific surface of powders.
4688:1971. Method for the determination of flash point (open) and fire point of petroleum products by the Pensky–Martens apparatus.
4726:1971. Methods for sampling raw materials for paints and varnishes.

Appendix C

COMPOSITION OF INORGANIC PIGMENTS AND EXTENDERS

(where composition is not apparent from the name)

Material	Composition
Antimony vermilion	Antimony sulphide
Asbestine	Natural calcium magnesium silicate
Barytes	Natural barium sulphate
Bauxite residue	Mainly oxide of iron
Black, Bone	Carbon mixed with calcium phosphate
Black, Carbon	Almost pure carbon
Black, Drop	The best quality Bone Black
Black, Gas	Identical with Carbon Black
Black, Lamp	Carbon (identical with Vegetable Black)
Blacklead	Identical with graphite
Black, Vegetable	Carbon (identical with Lamp Black)
Blanc Fixe	Precipitated barium sulphate
Blue, Bronze	Ferric ammonium (or sodium) ferrocyanide
Blue, Brunswick	Prussian Blue with extender
Blue, Celestial	Prussian Blue with extender
Blue, Cerulean	Mixed oxides of cobalt and tin
Blue, Chinese	Term formerly applied to best Prussian Blue
Blue, Cobalt	Mixed oxides of cobalt and aluminium
Blue, French	Identical with Ultramarine Blue
Blue, Milori	A variety of Prussian Blue
Blue, Prussian	Ferric potassium ferrocyanide
Blue, Thénard's	Identical with Cobalt Blue
Blue, Ultramarine	Complex sulphosilicate of aluminium and sodium
Brown, Purple	Iron oxide
Bronze powders	Alloys of copper
Cadmopone	Co-precipitated cadmium sulphide and barium sulphate
Calcite	Natural crystalline calcium carbonate
China clay	Natural hydrated aluminium silicate
Chrome, Lemon	Lead chromate and lead sulphate
Chrome, Mid	Lead chromate
Chrome, Orange	Basic lead chromate
Chrome, Primrose	Lead chromate, lead sulphate and alumina
Chrome, Scarlet	Lead chromate, lead molybdate and lead sulphate
Graphite	Semi-crystalline carbon
Green, Bronze	A chrome green modified with other pigments, often oxides of iron
Green, Brunswick ⎫ Green, Chrome ⎬	Mixtures of lead chrome and Prussian Blue
Green, Cobalt	Mixed oxides of cobalt and zinc
Green, Guignet's	Hydrated oxide of chromium
Gypsum	Natural hydrated calcium sulphate
Kaolin	Identical with china clay
Kieselguhr	A natural diatomaceous silica
Lapis lazuli	Natural form of Ultramarine Blue (q.v.)
Lead, Orange ⎫ Lead, Red ⎬	Oxide of lead (Pb_3O_4)

Material	Composition
Lead, White	Basic lead carbonate
Lithopone	Co-precipitated zinc sulphide and barium sulphate
Magnesite	Natural magnesium carbonate
Maroon, Cadmium	Co-precipitated cadmium sulphide and selenide
Mica	Natural hydrated aluminium potassium silicate
Micaceous oxide	Natural black iron oxide
Ochres	Hydrated oxides of iron and siliceous matter
Red, Cadmium	Co-precipitated cadmium sulphide and selenide
Red, Indian	Synthetic iron oxide
Red, Turkey	Synthetic iron oxide
Red, Venetian	Synthetic iron oxide with calcium sulphate
Sienna, Burnt	Iron oxide containing manganese oxide and siliceous matter
Sienna, Raw	Hydrated iron oxide, manganese oxide and siliceous matter
Talc	Natural hydrated magnesium silicate
Umber, Burnt	Iron oxide, manganese oxide and siliceous matter
Umber, Raw	Hydrated iron oxide, manganese oxide and siliceous matter
Veridian	Hydrated chromium oxide
Vermilion	Mercuric sulphide
Violet, Cobalt	Cobalt pyrophosphate
White, Antimony	Antimony oxide
White, Chinese	Zinc oxide
White, Flake	Basic lead carbonate (identical with white lead)
White, Mineral	Identical with Gypsum
White, Paris	Calcium carbonate
White, Satin	Calcium sulphate and aluminate
White, Strontium	Strontium sulphate
White, Zinc	Identical with Lithopone
Whiting	Calcium carbonate
Witherite	Natural barium carbonate
Yellow, Cadmium	Cadmium sulphide

Appendix D

TEMPERATURE CONVERSION TABLES

°C	°F	°C	°F	°C	°F	°C	°F
−17·8	0	22	71·6	70	158·0	128	262·4
−15·0	5	23	73·4	72	161·6	130	266·0
−12·2	10	24	75·2	74	165·2	132	269·6
− 9·4	15	25	77·0	76	168·8	134	273·2
− 6·6	20	26	78·8	78	172·4	136	276·8
− 4·0	25	27	80·6	80	176·0	138	280·4
− 1·1	30	28	82·4	82	179·6	140	284·0
0	32	29	84·2	84	183·2	142	287·6
1	33·8	30	86·0	86	186·8	144	291·2
2	35·6	31	87·8	88	190·4	146	294·8
3	37·4	32	89·6	90	194·0	148	298·4
4	39·2	34	93·2	92	197·6	150	302·0
5	41·0	36	96·8	94	201·2	160	320·0
6	42·8	38	100·4	96	204·8	170	338·0
7	44·6	40	104·0	98	208·4	180	356·0
8	46·4	42	107·6	100	212	190	374·0
9	48·2	44	111·2	102	215·6	200	392·0
10	50·0	46	114·8	104	219·2	210	410·0
11	51·8	48	118·4	106	222·8	220	428·0
12	53·6	50	122·0	108	226·4	230	446·0
13	55·4	52	125·6	110	230·0	240	464·0
14	57·2	54	129·2	112	233·6	250	482·0
15	59·0	56	132·8	114	237·2	260	500·0
16	60·8	58	136·4	116	240·8	270	518·0
17	62·6	60	140·0	118	244·4	280	536·0
18	64·4	62	143·6	120	248·0	290	554·0
19	66·2	64	147·2	122	251·6	300	572·0
20	68·0	66	150·8	124	255·2	310	590·0
21	69·8	68	154·4	126	258·8	315	600·0

Conversion factors

$$°C = (°F - 32) \times \frac{5}{9} \qquad\qquad °F = \frac{°C \times 9}{5} + 32$$

Appendix E

SOME COLOURING-MATTERS OF HISTORIC INTEREST

Aureolin A double nitrite of cobalt and potassium. A bright transparent yellow, very expensive but used occasionally as an artists' water colour.

Arsenic Yellow or King's Yellow Arsenic sulphide (As_2S_3). Originally the natural ore "orpiment" was used but later the pigment was manufactured. It was used as a yellow pigment before the advent of lead chromes.

Carmine Lake Produced from carminic acid (see Cochineal, below) by precipitation with calcium and aluminium salts. It possesses a brilliant red colour but poor lightfastness.

Cochineal A dyestuff extracted from the dried bodies of the female cochineal beetle. The principal component is carminic acid, an anthraquinone derivative.

Crimson Lake A lake precipitated from carminic acid by aluminium and tin salts. A bright colour and less fugitive than Carmine Lake.

Dutch Pink A yellow lake prepared from the extract of the bark of a North American oak. The colouring principle is quercitrin, a tetrahydroxyflavonol.

Egyptian Blue A pigment introduced in Egypt in about 3000 B.C. It remained the principal blue pigment in general use for many centuries; a crystalline silicate of copper.

Indigo (natural) A pigment obtained from various species of *Indigofera*. This has now been replaced by the synthetic material.

Logwood Prepared from the dye extracted from the wood of *Haematoxylin campechianum* (indigenous to Mexico and Central America). Brown to blue-black lakes are precipitated with various mordants.

Madder Lake The dyestuff from the root of *Rubia tinctorum*; contains alizarin and some purpurin. The lake is formed by treatment with aluminium salt and alkali. It shows good lightfastness and was formerly used as a standard for this property. The synthetic product is used as an artists' colour.

Persian Berries Lake A yellow lake prepared from an extract of the unripe berries of a species of buckthorn. The colouring principle is rhamnetin, a flavonol.

Purple Lake A lake produced in a manner similar to that for Carmine Lake (*above*) but with the addition of lime to produce the deep purple tone. It is used as an artists' colour.

Sepia A dark-brown pigment obtained from marine animals of the family Sepiidae (cuttle-fish). Its principal use was in artists' colours.

Tyrian Purple A bright pigment much used by some ancient peoples. It was obtained from a variety of snail, *Murex brandaris*. The colouring-matter is 6:6-dibromoindigotin.

Vandyke Brown A natural brown earth containing ferric oxide. It continues to be used as an artists' colour.

Weld A yellow lake prepared from an extract of *Reseda luteola*. The colouring principle is luteolin, a tetrahydroxyflavanone.

RESINS

The following list contains some resins of historic interest only; others are used on a small and diminishing scale.

Amber A fossil type and the hardest known resin. Used in some forms of jewellery.

Copals This term covers both fossil and "modern" types. The latter continue to be used, but the former, which are soluble in oils only after thermal treatment ("gum running"), have been largely replaced by the hard synthetic resins.

Dragon's blood A red-coloured resin exuded from a species of plant native to eastern Asia. At one time it was used for colouring lacquers.

Gamboge A yellow gum resin obtained from certain trees indigenous to Thailand. It has poor chemical resistance and lightfastness.

Kauri A hard fossil copal obtained from New Zealand. It is now very expensive and used only for special purposes.

Bibliography

BARRY, T. HEDLEY, *Natural Varnish Resins*. Ernest Benn, London, 1932.

Appendix F

CONVERSION FACTORS (× = multiply by)

Length

"Thou" or "mils" to micrometres (μm)	× 25·4
Inches to millimetres	× 25·4
Yards to metres	× 0·914
Miles to kilometres	× 1·61

Area

Square inches to square centimetres	× 6·4516
Square feet to square centimetres	× 929·03
Square yards to square metres	× 0·836

Volume

Pints to cubic centimetres	× 567·9
Pints to litres	× 0·568
Gallons (Imp.) to litres	× 4·54
Gallons (U.S.) to litres	× 3·79
Cubic yards to cubic metres	× 0·764

Weight

Ounces to grams	× 28·35
Pounds to kilograms	× 0·454
Cwts to kilograms	× 50·85
Tons to kilograms	× 1016

Density

Lbs/cu. foot to kg/cu. metre	× 16·0

RELATIVE ATOMIC WEIGHTS OF SOME COMMON ELEMENTS

Element	Symbol	Atomic weight
Aluminium	Al	26·97
Antimony	Sb	121·76
Arsenic	As	74·91
Barium	Ba	137·36
Boron	B	10·82
Bromine	Br	79·92
Cadmium	Cd	112·41
Calcium	Ca	40·08
Carbon	C	12·01
Cerium	Ce	140·13
Chlorine	Cl	35·46
Chromium	Cr	52·01
Cobalt	Co	58·94
Copper	Cu	63·57
Fluorine	F	19·00
Hydrogen	H	1·008
Iodine	I	126·92

Element	Symbol	Atomic weight
Iron	Fe	55·85
Lead	Pb	207·21
Lithium	Li	6·94
Magnesium	Mg	24·32
Manganese	Mn	54·93
Mercury	Hg	200·61
Molybdenum	Mo	95·95
Nickel	Ni	58·69
Nitrogen	N	14·01
Oxygen	O	16·00
Phosphorus	P	30·98
Potassium	K	39·10
Selenium	Se	78·96
Silicon	Si	28·06
Silver	Ag	107·88
Sodium	Na	23·00
Strontium	Sr	87·63
Sulphur	S	32·06
Tin	Sn	118·70
Titanium	Ti	47·90
Tungsten	W	183·90
Vanadium	V	50·95
Zinc	Zn	65·38
Zirconium	Zr	91·22

Subject index

Author index